Astrophysics of Red Supergiants

AAS | IOP Astronomy

About the program:

AAS-IOP Astronomy ebooks is the official book program of the American Astronomical Society (AAS), and aims to share in depth the most fascinating areas of astronomy, astrophysics, solar physics and planetary science. The program includes publications in the following topics:

- GALAXIES AND COSMOLOGY
- INTERSTELLAR MATTER AND THE LOCAL UNIVERSE
- STARS AND STELLAR PHYSICS
- EDUCATION, OUTREACH AND HERITAGE
- HIGH-ENERGY PHENOMENA AND FUNDAMENTAL PHYSICS
- THE SUN AND THE HELIOSPHERE
- THE SOLAR SYSTEM, EXOPLANETS, AND ASTROBIOLOGY
- INSTRUMENTATION, SOFTWARE, LABORATORY ASTROPHYSICS AND DATA

Books in the program range in level from short introductory texts on fast-moving areas, graduate and upper-level undergraduate textbooks, research monographs and practical handbooks.

For a complete list of published and forthcoming titles, please visit iopscience.org/books/aas.

About the American Astronomical Society

The American Astronomical Society (aas.org), established 1899, is the major organization of professional astronomers in North America. The membership (~7,000) also includes physicists, mathematicians, geologists, engineers and others whose research interests lie within the broad spectrum of subjects now comprising the contemporary astronomical sciences. The mission of the Society is to enhance and share humanity's scientific understanding of the universe.

Astrophysics of Red Supergiants

Emily M. Levesque
University of Washington

IOP Publishing, Bristol, UK

ISBN 978-0-7503-1329-2 (ebook)
ISBN 978-0-7503-1330-8 (print)
ISBN 978-0-7503-1331-5 (mobi)

DOI 10.1088/978-0-7503-1329-2

Version: 20171201

AAS–IOP Astronomy
ISSN 2514-3433 (online)
ISSN 2515-141X (print)

British Library Cataloguing-in-Publication Data: A catalogue record for this book is available from the British Library.

Published by IOP Publishing, wholly owned by The Institute of Physics, London

IOP Publishing, Temple Circus, Temple Way, Bristol, BS1 6HG, UK

US Office: IOP Publishing, Inc., 190 North Independence Mall West, Suite 601, Philadelphia, PA 19106, USA

For Dad, who (with the help of Betelgeuse and our backyard telescope) showed me my first red supergiant.

Contents

Acknowledgements

This book would not have been possible without the invaluable efforts of my colleagues in the fields of massive stars and stellar evolution. Particular gratitude is due to Philip Massey for his many years of mentorship and collaboration.

Many thanks as well to Henny Lamers for his feedback on the manuscript, and to Paula Szkody for suggesting that I write a book in the first place.

I could not have completed this book without the stellar energy and enthusiasm provided by the current and former members of my Massive Stars research group at the University of Washington, including Nell Byler, Brooke Dicenzo, Trevor Dorn-Wallenstein, Kristen Garofali, Jeffrey Jennings, Rubab Khan, Jamie Lomax, Kathryn Neugent, Nathen Nguyen, Locke Patton, Mallory Thorp, and Kolby Weisenburger.

Finally, much appreciation is due to Leigh Jenkins, Daniel Heatley, and everyone else at IOP Publishing for their tireless assistance and support.

Author Biography

Emily M. Levesque

Emily Levesque is an Assistant Professor of Astronomy at the University of Washington in Seattle. She received her S.B. in physics from the Massachusetts Institute of Technology in 2006 and her PhD in astronomy from the University of Hawaii in 2010. From 2010 to 2015 she was an Einstein Fellow and Hubble Fellow at the University of Colorado at Boulder. Her primary research area is massive stellar astrophysics, with a focus on evolved massive stars, such as red supergiants, and their potential role as cosmological tools. Her research accolades include the 2014 Annie Jump Cannon research prize awarded by the American Astronomical Society and a 2017 Alfred P. Sloan fellowship in physics.

AAS | IOP Astronomy

Astrophysics of Red Supergiants

Emily M. Levesque

Chapter 1

An Introduction to Red Supergiants

Massive stars are critical objects for multiple areas of astronomy. Luminous massive star populations dominate the radiation signatures of young star-forming galaxies. Massive stars also serve as lighthouses during their explosive deaths as supernovae, drawing our attention to sites of recent star formation and giving us fleeting glimpses of stars born in the earlier eras of our universe. These same populations of massive stars can also be studied in incredible detail in the Milky Way and its neighboring galaxies, which offer a treasure trove of nearby targets that can be used as local laboratories. For massive stars alone it is possible to detect early generations of these objects exploding in distant galaxies while also closely examining the physical properties of their evolutionary and chemical twins in our own cosmic backyard. We *must* develop a thorough and comprehensive understanding of massive stars and their evolution if we wish to understand the fundamentals of stellar physics, star-forming galaxies, supernovae, and the chemical enrichment history of the universe.

Red supergiants (RSGs) are the helium-fusing, evolved descendants of moderately massive (~8–40$M_{\odot}$) main-sequence stars, the end result of a nearly horizontal evolution across the top of the Hertzsprung–Russell diagram. They are the coldest and largest (in physical size) members of the massive star population and the most luminous cool stars in the universe, representing a crucial extreme of stellar evolution. During their time as RSGs they undergo an impressive amount of mass loss, contributing to the dust content and chemical enrichment of the interstellar medium. They are expected to be a critical stage in the formation of interacting binaries, overflowing their Roche lobes at large radii and kicking off mass=transfer and interactions with previously undisturbed gravitational companions. RSGs are also crucial stages in the deaths of massive stars, capable of dying spectacularly as Type II-P core-collapse supernovae or of winking out quietly as "disappearing" stars that may have directly collapsed to black holes. At higher masses they serve as a critical transition and mass-losing phase for stars that eventually die in a later phase of evolution.

doi:10.1088/978-0-7503-1329-2ch1

In their classic 1990 textbook *Stellar Structure and Evolution*, Rudolf Kippenhahn and Alfred Weigert described the evolved stages of massive stars as a "sort of magnifying glass, revealing relentlessly the faults of calculations of earlier phases." RSGs are one of the most invaluable magnifying glasses we have for studying massive stars and their role in the universe. These stars have enormous relevance to a broad range of current fields, as well as to the next generation of observing facilities.

1.1 A Key Evolutionary Phase

For a star to be considered an RSG it must, by definition, have had an initial mass of at least $\sim 8M_{\odot}$, marking it as a massive star. Prior to the start of the RSG stage it has spent $\sim 10^6$–10^7 years on the main sequence, fusing hydrogen primarily via the carbon-nitrogen-oxygen cycle. When this phase is complete and the supply of hydrogen in the core has been exhausted, the stars streak across the top of the Hertzsprung–Russell diagram in an extremely short amount of time ($\sim 10^3$–10^5 years) before becoming RSGs, settling at cool effective temperatures ($\sim$3500–4500 K) and large radii (~ 100–$1000R_{\odot}$) as core helium fusion begins. In this way red *super*giants are quite distinct from red *giants*. Their higher initial masses, shorter main-sequence lifetimes, and luminosity and temperature evolution prior to the start of their red phase are all markedly different from the evolution of low-mass stars, which spend billions of years fusing hydrogen before progressing through the red giant branch, horizontal branch, and asymptotic giant branch (AGB) after leaving the main sequence, experiencing multiple dredge-up events during this latter phase that bring fusion products to the star's surface. Their subsequent evolution is also drastically different: RSGs, like other massive stars, will proceed through later stages of core fusion (C fusion and beyond) before undergoing core collapse, while the vast majority of red giants and AGB stars produce white dwarfs and planetary nebulae following the end of hydrogen and helium fusion. While there is some observational ambiguity in distinguishing an extremely luminous AGB star from a relatively low-mass RSG (see Chapters 3 and 6), red giants and supergiants are distinct and very different stellar populations and should be treated as such (unfortunately similar naming conventions aside).

The criteria for calling a star an RSG are primarily observational—the designation is given to cool K-type and M-type supergiants, while the earlier and warmer F and G spectral types correspond to the fleeting yellow supergiant phase that precedes (and, in some cases, also follows) the RSG phase. An even simpler set of criteria applies cutoffs in absolute magnitude and color index, literally demarking RSGs as red and bright (presumably massive) stars in the upper right corner of the color–magnitude diagram. Translating these criteria from the (observational) color-magnitude diagram to the (theoretical) Hertzsprung–Russell diagram illustrates the corresponding physical criteria, with effective temperatures of $\lesssim$4500 K and bolometric luminosities of $\gtrsim 10^4 L_{\odot}$ in the Milky Way. From an evolutionary standpoint these physical properties correspond to core helium fusion (and in later stages, for some masses, core C fusion) and an extended fully convective envelope, although in some scenarios (e.g., extreme metallicities, strong mass loss, or interacting binary evolution) the mapping between observational traits and evolutionary states can be more complicated.

Stars during this phase represent vital extremes in the lifetimes of massive stars. Massive stars lose more than half of their mass after they evolve off of the main sequence, and a significant fraction of this mass loss occurs during the RSG phase. The physical mechanisms that contribute to and dominate this mass loss are still unclear (see Chapter 4), but observational evidence has demonstrated that this mass loss contributes to RSGs producing substantial amounts of dust, potentially dominating dust production in young stellar populations. This mass loss also contributes strongly to chemical enrichment, produces a morphologically complex circumstellar environment that is sometimes later evidenced during the star's death as a core-collapse supernova, and is a key step in how more massive stars eventually exit the RSG phase by again evolving leftward on the Hertzsprung–Russell diagram (mass loss that strips away the outer layers is one explanation for these stars' eventual evolution to post-RSG yellow supergiants and/or Wolf–Rayet stars). RSGs also represent the physically largest evolutionary phase possible for single stars, making them critical ingredients for understanding the total fractions, mass ranges, and evolutionary states of interacting massive binaries due to the large interaction radii that they can achieve with their companions through Roche lobe overflow. This interaction can lead to mass-transfer-induced stripping of the star's outer layers, massive common envelope binaries, and even, in rare cases, stellar mergers that can produce exotic phenomena such as Thorne–Żytkow objects.

1.2 Supernova Progenitors

Stars end their time as RSGs in one of two ways: lower-mass RSGs are expected to undergo core collapse, while more massive RSGs evolve blueward on the Hertzsprung–Russell diagram and enter a new evolutionary phase.

Low-mass RSGs ($\lesssim 17 M_\odot$; Smartt 2015) currently dominate the (admittedly small) sample of observationally confirmed supernova progenitors, with about a dozen stars definitively detected in pre-explosion images of supernova sites and several other potential candidate RSGs (see Chapter 8). They are overwhelmingly associated with the Type II-P subclass of supernovae and are the presumed progenitors of this class as a whole—the "II-P" designation marks these as hydrogen-rich supernovae with light curves that show a distinct plateau phase in their evolution associated with the presence of a large hydrogen envelope in the progenitor, in excellent agreement with what one would predict from these lower-mass RSGs. At the same time, several mysteries still remain regarding the core-collapse deaths of these stars. RSGs at higher masses have notably *not* been detected as supernovae progenitors, and the explanation for this is still unclear. The stars could be difficult to capture in current imaging campaigns due to substantial circumstellar dust, their supernovae could be going undetected by current surveys, or the stars could be undergoing direct core collapse to black holes with no associated supernovae. These RSGs could also simply be evolving back to warmer temperatures and becoming yellow supergiants or Wolf–Rayet stars prior to exploding as supernovae.

This latter scenario is expected for high-mass RSGs, although the precise turning point in mass that dictates whether a star explodes as an RSG or evolves to later stages is unclear and dependent upon a wide range of other parameters (including metallicity, binary evolution, mass loss, and initial rotation rate). For the stars that do make this transition, they are thought to shift swiftly back across the top of the Hertzsprung–Russell diagram, briefly becoming yellow supergiants again before entering the Wolf–Rayet phase, an extremely hot post-main-sequence evolutionary phase of massive stars. It is worth noting that determining when a star has left the RSG phase is dictated primarily by surface and observational effects—extreme mass loss and color—rather than interior behavior (i.e., the star could be fusing He, C, or even heavier elements during this evolution). At the same time, understanding the correlation between the star's external and internal evolution during this phase is crucial for understanding the role that RSGs play as interim phases in the formation of hot post-main-sequence massive stars, and in turn *their* eventual role as supernova progenitors. Massive stellar evolutionary theory predicts that these stars will also produce other types of core-collapse supernovae, and both yellow supergiants and Wolf–Rayet stars have been at least tentatively observationally identified as supernova prognenitors (see Chapter 8 for further discussion). The mass loss and evolution of these progenitor stars and the circumstellar environments during the RSG phase will heavily influence the final evolutionary stage of these stars and the properties of the supernovae that they produce.

1.3 Extragalactic Probes

With their extremely high luminosities ($\gtrsim 10^4 L_\odot$) and bright optical and near-infrared magnitudes, RSGs are ideal targets for observations of individual stars in galaxies beyond our own Milky Way. Using current facilities, single RSGs can be detected in imaging reaching out to tens of Mpc, and moderate-resolution spectroscopy of individual RSGs has been achieved out to ~3Mpc—these distances allow us to stretch studies of extragalactic stellar astrophysics well beyond the Local Group.

The ability to study RSGs in other galaxies makes these stars particularly excellent "magnifying glasses" for stellar evolutionary theory. With their young ages, the metallicities of massive stars match the metallicities of their host environments. Observations of RSG populations in other galaxies with a diverse range of abundances thus give us the chance to study how massive stellar evolution is impacted by metallicity, directly testing the predictions of existing theory and providing new samples that can serve as a framework for developing future models. This application of RSG observations is particularly valuable at low metallicities, which still present a number of challenges to evolutionary theory—studying the galaxies of the early universe is critically dependent on our ability to accurately model their metal-poor massive star populations, which in turn requires a rigorous understanding of massive stellar evolution in these environments.

RSGs can also be used as metallicity probes in their own right. Current observational studies of star-forming galaxies and HII regions most often depend on metallicity diagnostics that use the ratios of strong emission lines in the galaxy spectra as a proxy for the gas phase metallicity. These methods are fraught with problems—most diagnostics have been developed from a hodgepodge of observational and theoretical resources (which are often also strongly dependent on massive stellar evolutionary theory, such as models developed from stellar population synthesis and photoionization codes) and disagree with each other to an alarming degree. Stellar metallicities, where possible, are an excellent and reliable means of measuring metallicity in other galaxies, and RSGs are ideal targets for this technique both as individual stars and as the dominant members of young stellar clusters. Their utility as metallicity indicators has already been demonstrated in galaxies both in and beyond the Local Group.

The extragalactic applications of RSGs will be particularly crucial in the era that is about to be ushered in by the next generation of telescopes. All of the upcoming Extremely Large Telescopes will have primary mirror diameters exceeding 20m and spectrographs and adaptive optics systems as first light instruments, dramatically expanding the volume where images and spectra of individual RSGs are possible. The James Webb Space Telescope (JWST) and the Wide Field Infrared Survey Telescope (WFIRST), both optimized as IR instruments, will also be capable of capturing individual RSGs at exceptional distances and studying the more nearby populations of RSGs in the Local Group and other nearby galaxies in incredible detail.

It is clear that RSGs occupy a unique cross-disciplinary niche as a key phase of massive stellar evolution, supernova progenitors, and potential probes of extragalactic stellar populations. However, while many researchers (in subfields ranging from stellar astrophysics to supernovae and high-redshift galaxies) are interested in some aspect of RSGs, there is no single reference that can be used to provide an effective overview of RSGs' fundamental physical properties, evolution, and their extragalactic and cosmological applications. The goal of this book is to serve as a compendium of our current knowledge, providing a one-stop reference across multiple fields for people seeking to learn more about RSGs.

Reference

Smartt, S. 2015, PASA, **32,** 16

Chapter 2

Inside a Red Supergiant

2.1 Evolution and the Stellar Core

Prior to becoming an RSG, a massive star has progressed from the zero-age main sequence to the terminal-age main sequence (TAMS), fusing hydrogen in its core predominantly via the CNO cycle. Adopting a simple example for a pre-RSG massive star (a $15M_{\odot}$ single non-rotating star with solar metallicity; see Figure 2.1), the star is in hydrostatic and radiative equilibrium and will become slightly more luminous during its time on the main sequence as a result of the μ-effect, where the mean particle mass μ in the core increases as H is converted into He. Deriving the mass-luminosity relation for such a star demonstrates that $L \approx \frac{\mu^4}{\kappa} \cdot M$ (assuming a homogenous composition with the opacity κ), so $L \propto \mu^4$ (in reality, for a non-homogenous interior the relation is slightly less steep), leading to an increase in L. The stellar radius also expands during the main sequence, a response to a decrease in central pressure while in hydrostatic equilibrium, leading to a corresponding slight decrease in T_{eff} as a result of $L \propto R^2 T_{\mathrm{eff}}^4$. This basic main-sequence evolution is consistent across a wide range of stellar models—varying treatments of metallicity, differential rotation, and convective overshooting mainly impact the lifetime of the star on the main sequence (~11Myr for our example star; Ekström et al. 2012) and its corresponding luminosity when it reaches the TAMS.

This luminosity remains almost constant after the TAMS. Since the core is convective and therefore well-mixed, H is exhausted almost simultaneously throughout the core. The abrupt ceasing of H fusion leads to the core (and the entire star) contracting, producing a small leftward hook in the evolutionary track. This contraction halts when the shell of material directly outside the former core reaches the temperature and density required for H fusion. The interior of our star is now briefy undergoing H fusion in a low-mass shell around the core, which continues to rapidly contract in pursuit of achieving the necessary temperature and density for core He fusion. At this point most of the mass of the star is still in radiative

doi:10.1088/978-0-7503-1329-2ch2

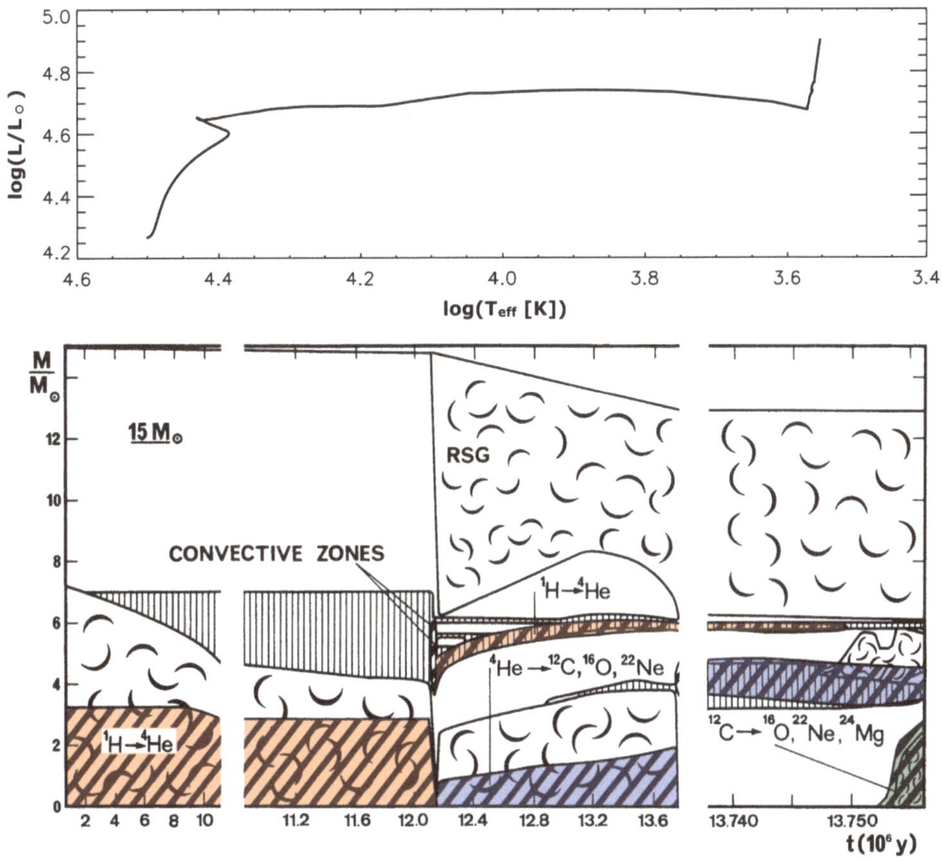

Figure 2.1. Top: evolutionary track of a non-rotating solar-metallicity $15M_{\odot}$ star, as computed in Ekström et al. (2012). Bottom: adapted from Maeder & Meynet (1987) and Lamers & Levesque (2017); Kippenhahn diagram for a $15M_{\odot}$ star at solar metallicity, including zones of H fusion (red), He fusion (blue), and C fusion (green); regions where the energy generation rate $\varepsilon > 10^3$ erg g^{-1} s^{-1} (indicated by diagonals); convective regions (curls); regions of variable H and He (vertical lines); and regions of variable C, O, and Ne (horizontal lines).

and hydrostatic equilibrium (note that, unlike their lower-mass cool giant counterparts, RSGs do not develop degenerate cores), so the star is still obeying the mass-luminosity relation: with μ, κ, and M all nearly constant, L likewise varies little during H shell fusion.

At the same time, the combination of the newly ignited H-fusion shell and contracting core produces the famed "mirror effect," leading to an expanding stellar envelope (while still not well-understood, this behavior is most often explained by invoking the virial theorem or pressure effects, which describe the opposing actions of the core and envelope as a means of maintaining the temperature, and thus the energy generation rate, of the fusion shell). With L held constant and R increasing rapidly, $T_{\rm eff}$ now drops dramatically. The result is an extremely rapid and almost perfectly horizontal rightward evolution of our star on the Hertzsprung–Russell diagram (HRD); it briefly becomes a yellow supergiant as it sprints across the HRD in just $\sim 10^4$ years. This rapid horizontal evolution produces the "yellow gap" phenomenon on the HRD. Most massive stars spend relatively little time in this intermediate temperature region, leading to short yellow supergiant lifetimes and a resulting very small number of these stars in massive stellar populations.

By the time core He fusion begins, halting the contraction of the convective core, our star has almost the same L it left the TAMS with but a much larger R ($\sim 100 R_\odot$ or more) and a much cooler $T_{\rm eff}$ (~3500–4000 K), beginning its time as an RSG.

Early in the RSG phase the star is fusing He (producing C and O) in its core via the triple-α process along with H in a shell; this phase lasts ~1-2 Myr for our example $15M_\odot$ star (Ekström et al. 2012), though the expected duration can vary depending on treatments of parameters such as mixing (through convective overshooting or differential rotation), the μ gradient, and the star's initial composition (the effects of convective overshooting, which in general increase the core mass fraction and lengthen the duration of the fusion phase, are less drastic for core He fusion compared to core H fusion due to a steeper pressure gradient as a function of mass; Maeder & Meynet 1987). As seen during core H fusion on the main sequence, μ increases and yields a corresponding increase in L, leading to the characteristic upturn and vertical evolution of stellar evolutionary tracks on the HRD during the RSG phase, as shown in Figure 2.1 (however, unlike the main sequence, $T_{\rm eff}$ does *not* decrease; see Section 3.2).

Some of the lowest-mass RSGs ($\lesssim 12M_\odot$) may temporarily exit the RSG phase during this period of core He and shell H fusion, evolving to the left on the HRD and producing a "blue loop" in the evolutionary track as they temporarily return to a yellow supergiant state. While the underlying physics driving this loop is uncertain, one explanation invokes the same phenomenon that produces the "horizontal branch" in the evolution of lower-mass stars: expansion of the He-fusion core inside the H-fusion shell leads to a corresponding contraction of the envelope, yielding the opposite "mirror effect" result of that seen when the star first crosses the yellow gap at the start of shell H fusion. This behavior is dictated by the surface gravitational potential of the core inside the H fusion shell, Φ_c, which correlates with the core mass and radius as $\Phi_c \sim M_c/R_c$; if this potential is below some critical value, blue loops are produced. The exact critical value is unknown, but models have demonstrated that blue loops are generally suppressed at higher masses; under hydrostatic equilibrium, the core temperature $T_c \sim \Phi_c$, and with higher initial masses corresponding to higher T_c, Φ_c quickly exceeds the critical maximum for producing blue loops. The effects of a lowered opacity and increased mean molecular weight throughout the star's envelope (reducing the fuel supply in the fusion shell) can also favor blueward evolution (e.g., Walmswell et al. 2015). As with most other evolutionary stages, the expected onset, duration, and mass range of blue loops all vary with treatments of convective overshooting, rotation, mixing, metallicity, and even mass-loss rates (which can increase the homogeneity of higher-mass stars during core He fusion and may actually induce blue loop behavior at higher RSG masses; Meynet et al. 2015). Blue loops that extend to warm enough $T_{\rm eff}$ pass through the instability strip on the HRD, producing pulsations in these stars and making them the most massive and luminous members of the classical (Type I) Cepheid population.

In addition to core and shell fusion, RSGs also host the "weak" *s*-process (occurring with a lower neutron exposure than the main *s*-process), which contributes strongly to the abundance of elements with atomic numbers (A) between $60 < A < 90$ in the Milky Way and is expected to reach peak efficiency in $\sim$15-30$M_\odot$ stars beginning at the start of core He fusion (e.g., Raiteri et al. 1993; The et al. 2000, 2007). The efficiency and yield of this *s*-process, however, is strongly dependent on factors such as metallicity; the size of the convective core; mixing from convective overshooting and rotation; and the triple-α, ^{22}Ne(α,n)^{25}Mg, and ^{12}C(α,γ)^{16}O reaction rates (e.g., Tur et al. 2007, 2009; West et al. 2013; Jones et al. 2015). The weak *s*-process is also expected to occur primarily in the most central regions of the star; however, convection in RSGs reaches deep into the interior layers (see Section 2.2), and, combined with mixing, should bring the products of this nucleosynthesis to the surface.

The edges of the convective core and the H fusion shell both migrate outward during the core He fusion phase of an RSG, a consequence of the small and short-lived intermediate convection zone (Maeder & Meynet 1987). Eventually our massive star transitions to the later (and even shorter-lived) fusion phases, with H and He fusion shells surrounding a convective C-fusing core. The progress through C fusion and later core fusion phases produces the famous "onion-skin" cartoon interior model of a pre-supernova massive star that ultimately concludes with a Fe core and the star's subsequent supernova death (see Chapter 7), although effects such as mass loss and mixing will complicate this interior structure. It is also important to note that this interior evolution is (somewhat) decoupled from the outward appearance of the star. At lower masses stars are expected to complete this entire process as RSGs in the upper right of the HRD, while more massive RSGs will eventually evolve back to the left on the HRD due to the stripping away of their outermost layers by mass loss and the resulting exposure of their hotter interior layers, turning them into blue supergiants or Wolf–Rayet stars (see Chapters 3 and 8).

The initial masses and subsequent interior structures and evolutionary progressions of RSGs distinguish them from red giants and asymptotic giant branch (AGB) stars in several key ways. With their lower initial masses ($\lesssim 8 M_\odot$), red giants and AGB stars have much longer main core H fusion lifetimes and can attribute their position as luminous red stars on the HRD to very different interior physics. While RSGs climb vertically during core He fusion as a consequence of the μ-effect, the vertical evolution of giants along their parallel red branches is driven by the combination of shell fusion and a degenerate He (red giants) or CO (AGB stars) core. As shell fusion progresses, mass is deposited onto the degenerate core, causing it to shrink and producing an expanding radius via the mirror effect, which manifests as an increase in L rather than a decrease in $T_{\rm eff}$ (as seen in the post-TAMS evolution of RSG progenitors) due to the stars' placement at the Hayashi limit (the lower limit of $T_{\rm eff}$ for stars in hydrostatic equilibrium; see Section 3.2). This effect also results in the convective envelope of the star deepening, reaching as far as the intershell region in AGB stars and producing the characteristic multiple dredge-ups seen in these lower-mass stars that bring products of H and He fusion and the main *s*-process to the stellar surface and reduce both the He core and CO core mass, yielding a distinctly different interior structure with two fusion shells in close proximity (as opposed to the simpler RSG structure comprised of successive core and single-shell fusion stages).

In general, RSGs are also distinguished from their lower-mass red counterparts by the latter's inability to fuse elements beyond C (and their subsequent deaths as planetary nebulae and white dwarfs rather than supernovae). However, "super"-AGB stars—stars with initial masses ranging from ~7–d9$M_\odot$—represent an interesting middle ground. While their evolution outwardly bears more resemblance to their low-mass counterparts, with at least partially degenerate cores and thermal pulses characteristic of the AGB phase, they are capable of fusing C and may, in the most massive cases, be capable of producing core collapse supernovae (e.g., Poelarends et al. 2008). Super-AGB stars are also extremely luminous, and observationally it can be difficult to distinguish them from low-mass RSGs on the HRD (see further discussion in Chapters 3 and 6).

The exterior and interior evolution of our example 15$M_\odot$ star is summarized in Figure 2.1; we can follow its outer evolution on the HRD (top) and the changes in its interior via the Kippenhahn diagram (which illustrates the evolution of a star's internal structure as a function of time on the x-axis; bottom); in both cases the star is modeled up through core C fusion.

2.2 Envelopes and Atmospheres

The outer layers of RSGs are convective, a consequence of their large radii, cool temperatures, and high opacities (due to the partial ionization of H^-). Combined with the convective core and shell fusion regions, these features make RSGs almost completely convective from core to surface.

Schwarzschild (1975) was the first to propose that RSG atmospheres should be dominated by a small number of extremely large convective cells, with characteristic sizes on the order of the stellar radius itself. This was in agreement with the predictions of mixing length theory (Böhm-Vitense 1958), which states that the characteristic scale of convection is dictated by the atmospheric pressure scale height, H_p. Schwarzschild's prediction was that an RSG should, at its surface, be covered by about a dozen convection cells with timescales dictated by the size of the convective element and the sound speed c_s (~5 km s^{-1} at the surface of an RSG). This was later supported theoretically by Freytag et al. (2002) and observationally by Josselin & Plez (2007) based on velocity variations in high-resolution spectra of Milky Way RSGs.

Chiavassa et al. (2009) also noted that reproducing the large sizes of convection cells depends on a careful treatment of H_p in RSGs that takes into account the stars' high turbulent pressure, which can dominate the gas pressure in turbulent convective layers. This modifies H_p from the standard definition, where $H_p = \frac{kT_{\mathrm{eff}}}{g\mu m_H}$, to one accounting for the turbulent pressure where $H_p = \frac{kT_{\mathrm{eff}}}{g\mu m_H}(1 + \beta\gamma(\frac{v_{\mathrm{turb}}}{c_s})^2)$. The second term in the parentheses includes the turbulent velocity $v_{\rm turb}$, the sound speed c_s, the adiabatic exponent γ, and the parameter β whose value depends on the anisotropy of the velocity field, and considerably increases the pressure scale height (see also Freytag et al. 1997; Gustafsson et al. 2008). This treatment of the convection cell size shows good agreement with observations; Chiavassa et al. (2010) measured the characteristic sizes of convective cells on Betelgeuse using IR observations and found them to be in good agreement with model predictions.

Chiavassa et al. (2009, 2011) produced three-dimension radiative hydrodynamical simulations of convection in RSGs, simulating the evolution of these large convective cells in the stars' outer layers and modeling the effect that this behavior should have, observationally, on RSG surfaces (see Figure 2.2). Their models produced convective cells with strong downdrafts, capable of penetrating all the way to the (also convective) stellar core (e.g., Freytag et al. 2002). The convective envelopes of RSGs also produce acoustic waves, a consequence of supersonic (five times the sound speed or higher) non-stationary convective flows; these waves travel outward into the convective envelope and even into the atmosphere, where they are compressed and amplified, producing shocks (see Chiavassa & Freytag 2015) and complicating standard treatments of mixing length theory and local thermodynamic equilibrium.

The mechanical coupling of the stellar layers inside a RSG should also be noted here. The expansion in radius before and during the RSG phase leads to a significant drop in the surface rotation rate of the star (see Chapter 3), but how that translates to the inner layers (including the core, whose rotation rate becomes very important during core collapse and the formation of a compact object) is unclear. The coupling of the surface through to the core of the star depends strongly on the degree of mixing between the stellar layers and the treatment of interior magnetic fields. A high degree of differential rotation in an RSG can potentially leave the core and innermost layers almost untouched, rotating much faster than the surface and losing

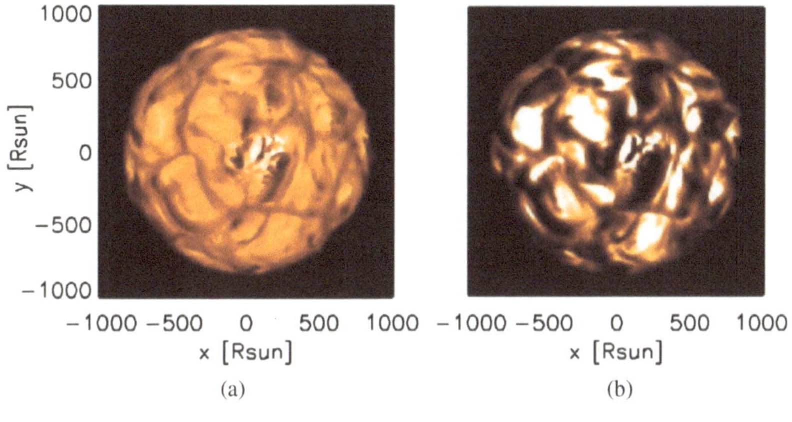

Figure 2.2. Reproduced with permission from Chiavassa et al. (2009, 2011; © ESO.); animated movies of synthetic linear intensity maps in the H band (a) and TiO band (b), illustrating the expected surface appearance of convective cells in the stellar envelope. The star modeled is an RSG with an initial mass of $12M_{\odot}$, $L = 90000L_{\odot}$, $T_{\rm eff} = 3500$ K, and $R = 832R_{\odot}$. Animation available from https://doi.org/10.1088/978-0-7503-1329-2.

relatively little angular momentum. Interestingly, differential rotation between a contracting core and outer envelope is also a potential means of producing the magnetic fields that have been observed in some RSGs and invoked as a potential driver of mass loss (Uchida & Bappu 1982; see Chapter 4).

Combined, these phenomena make the outer layers and atmospheres of RSGs extremely difficult to simulate. However, current model codes such as MARCS (e.g., Plez 1992; Gustafsson et al. 2008) and PHOENIX (e.g., Lancon et al. 2007; Husser et al. 2013) are able to produce good agreement with the observed spectral features of RSGs in the optical and IR. This is in spite of the computationally necessary approximations that must be made when calculating these large grids of stellar atmosphere models; the current versions of both codes have calculated one-dimensional spherically symmetric models that assume local thermodynamic equilibrium and use mixing length theory to describe convection. Despite the acknowledged limitations, these models offer what are currently the best theoretical tools available for interpreting observations of RSGs—their spectra in particular—and determining the physical properties that allow us to place these stars in context with respect to the general population of cool and massive stars.

References

Böhm-Vitense, E. 1958, ZAp, **46, 108**

Chiavassa, A., & Freytag, B. 2015, in ASP Conf. Ser. 497, Why Galaxies Care about AGB Stars III: A Closer Look in Space and Time, F. Kerschbaum, R F. Wing, & J. Hron (San Francisco, CA: ASP), 11

Chiavassa, A., Haubois, X., Young, J. S., et al. 2010, A&A, **515, 12**

Chiavassa, A., Pasquato, E., Jorissen, A., et al. 2011, A&A, **528, 120**

Chiavassa, A., Plez, B., Josselin, E., & Freytag, B. 2009, A&A, **506, 1351**

Davies, B., Kudritzki, R.-P., Plez, B., et al. 2013, ApJ, **767, 3**

Ekström S., Georgy, C., Eggenberger, P., et al. 2012, A&A, **537, 146**

Freytag, B., Holweger, H., Steffen, M., & Ludwig, H.-G. 1997, in ESO Astrophysics Symp., Science with the VLT Interferometer (Berlin: Springer), **316**

Freytag, B., Steffen, M., & Dorch, B. 2002, AN, **323, 213**

Gustafsson, B., Edvardsson, B., Eriksson, K., et al. 2008, A&A, **486, 951**

Husser, T.-O., Wende-von Berg S., Dreizler, S., et al. 2013, A&A, **553, 6**

Jones, S., Hirschi, R., Pignatari, M., et al. 2015, MNRAS, **447, 3115**

Josselin, E., & Plez, B. 2007, A&A, **469, 671**

Lamers, H. J. G. L. M., & Levesque, E. M. 2017, Understanding Stellar Evolution, (Bristol: IOP)

Lancon, A., Hauschildt, P. H., Ladjal, D., & Mouhcine, M. 2007, A&A, **468, 205**

Levesque, E. M., Massey, P., Olsen, K. A. G., et al. 2005, ApJ, **628, 973**

Maeder, A. & Meynet, G., 1987, A&A, **182, 243**

Meynet, G., Chomienne, V., Ekström S., et al. 2015, A&A, **575, 60**

Plez, B. 1992, A&AS, **94, 527**

Poelarends, A. J. T., Herwig, F., Langer, N., & Heger, A. 2008, ApJ, **675, 614**

Raiteri, C. M., Gallino, R., Busso, M., et al. 1993, ApJ, **419, 207**

Schwarzschild, M. 1975, ApJ, **195, 137**

The, L.-S., El, Eid M. F., & Meyer, B. S. 2000, ApJ, **533, 998**

The, L.-S., El, Eid, M. F., & Meyer, B. S. 2007, ApJ, **655, 1058**
Tur, C., Heger, A., & Austin, S. M. 2007, ApJ, **671, 821**
Tur, C., Heger, A., & Austin, S. M. 2009, ApJ, **702, 1068**
Uchida, Y., & Bappu, M. K. V. 1982, ApA, **3, 277**
Walmswell, J. J., Tout, C. A., & Eldridge, J. J. 2015, MNRAS, **447, 2951**
West, C., Heger, A., & Austin, S. M. 2013, ApJ, **769, 2**

Emily M. Levesque

Chapter 3

Physical Properties of Red Supergiants

3.1 The Spectrum of a Red Supergiant

At first glance we expect the spectrum of an RSG to be, quite simply, *red*—with temperatures ranging from ~3400 to 4300K in the Milky Way the spectra should peak around 6750–8550Å according to the principles of blackbody radiation. In practice, however, the progression of the peak wavelength is difficult to discern (and to use as a temperature gauge) due to the increasing impact of line blanketing at later spectral types. As a result, the spectral types of RSGs have traditionally been determined based on the strength of several key features in the stars' optical spectra, most notably the broad absorption bands of TiO (see Figure 3.1).

TiO bands begin to appear in RSG spectra at a spectral type of K5; the 5167Å and 6158Å bands are the first to emerge, followed by the 4954Å and 7054Å bands as we shift to the M spectral type. As we move to later M types additional TiO bands begin to appear (most notably 4761Å, 5847Å, and 6658Å) and all of the bands increase in strength, eventually dominating the SED and completely blanketing out the original continuum. The strengths of these TiO bands are the primary means of determining RSG spectral types at K5 and later, and can also be used in conjunction with stellar atmosphere models to determine effective temperatures (see below for further discussion).

By contrast, early K-type RSGs have optical spectral energy distributions that are red and largely featureless, with few spectral lines that evolve with spectral type. The Balmer absorption features, which serve as powerful spectral type indicators in hotter stars, are relatively weak in RSG spectra (though Hα is valuable as a luminosity class diagnostic; see Section 3.6), the Ca II triplet at 8498Å,8542Å,8662Å is present and diminishing but does not significantly evolve from the early-K to mid-M spectral types, and the G-band of CH at 4300Å is present but weak, disappearing by mid-K spectral types. The primary features cited by Jaschek & Jaschek (1990) as spectral type indicators (such as the ratios of Ti I 3999Å/Fe I 4005Å, Fe I 4144Å/Hδ, and Ca I 4226Å/Fe I 4250Å) are weak features in the far blue, where most RSGs are extremely

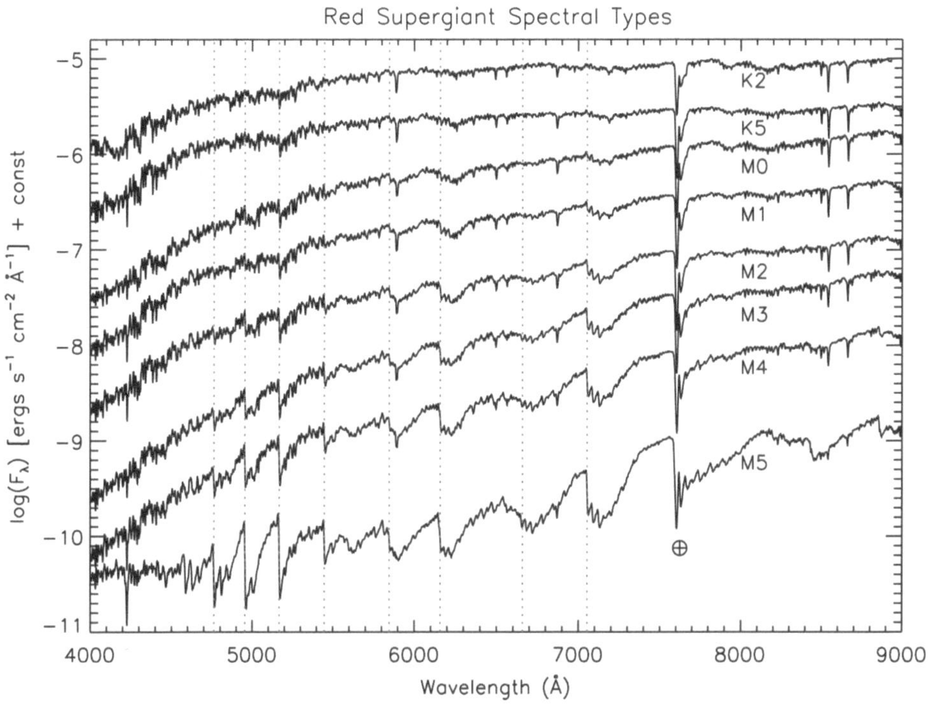

Figure 3.1. Spectral energy distributions of Milky Way RSGs from Levesque et al. (2005) spanning the full range of RSG spectral types from early K (top) to late M (bottom). The key TiO bands used for determining spectral types (4761Å, 4954Å, 5167Å, 5448Å, 5847Å, 6158Å, 6658Å, and 7054Å) are marked with gray dotted lines; the atmospheric O_2 A-band is marked with a ⊕.

dim—with lower-resolution spectra and observations of RSGs beyond the Milky Way, the low S/N in the blue renders that regime of the spectrum minimally useful for determining spectral types. Currently, the primary spectral type diagnostics for K supergiants are the weak CH G-band feature in the blue and the overall shape of the continuum; however, the continuum shape is strongly degenerate with the amount of circumstellar dust surrounding the RSG (see Section 4.2). In general, it is extremely difficult to determine spectral types for most K-type RSGs in the optical regime beyond very broad "early K" and "late K" classifications.

Beyond the optical regime, RSGs have best been studied in the near-IR (although no explicit spectral typing tools outside of the optical have been defined). As in the optical, their near-IR spectra are characterized by molecular bands, most notably the CN band at 1.1μm and the first and second overtone bands of CO at 2.3μm and 1.6μm, respectively (the VO and H_2O bands seen in stars with higher surface gravities are generally absent in RSG spectra; Lancon et al. 2007). The "H-hump" at 1.6μm, as named by Davies et al. (2013), also appears in some RSG spectra as a consequence of a H^- opacity minimum, but is suppressed at low surface gravities, low temperatures, and/or high metallicity by CO and CN molecular absorption.

In both the optical and the near-IR, RSG spectra contain a host of narrow atomic absorption lines, predominantly Fe I and Ti I. While these lines are not generally used for determining spectral types, they have been explored as potential effective temperature indicators in the optical (e.g., Dorda et al. 2016), though the strong relation between these lines and effective temperatures in F- and G-type stars breaks down in the K and M regime. More promisingly, near-IR atomic lines have been identified as useful diagnostics for determining stellar metallicities from RSGs in the near-IR (see Gazak et al. 2014 and Section 9.2).

At longer wavelengths, RSGs that have undergone significant mass loss can sometimes be identified in mid-IR colors, which can highlight circumstellar material produced by episodic mass-loss events (e.g., Britavskiy et al. 2014). While RSGs are not widely identified as significant radio sources, Dougherty et al. (2010) detected radio emission from all four RSGs in the Galactic cluster Westerlund 1; the emission was associated with both the central stars and their surrounding nebulae, which were also found to be ionized by potential hot companions (see Chapters 4 and 5 for further discussion). Some RSGs are significant maser sources in the submillimeter and radio, hosting circumstellar SiO, H_2O, and OH masers produced in the outflowing material generated by episodes of high mass loss—while this was previously associated with only the most luminous Galactic RSGs, more recent work suggests that ~15% of all Galactic RSGs show some associated maser activity (e.g., Davies et al. 2008; Menten et al. 2008; Verheyen et al. 2012; see also Section 4.3).

With a relative lack of flux compared to optical and near-IR observations, single RSGs are rarely studied in the ultraviolet (UV). However, eclipsing binary systems such as VV Cephei have been extensively observed in the UV, capturing the absorption spectrum produced when the B-type main-sequence companion emerges from eclipse and its UV continuum propagates through the M supergiant's atmosphere (e.g., Bauer et al. 2007; for more discussion of VV Cep, see Section 5.2). RSGs with hot companions can be identified by a significant upturn in their blue (<4000Å) and UV spectra, when the companion star's spectrum begins to dominate the SED. Finally RSGs that show signs of substantial circumstellar extinction in the visual also show significant excesses in their near-UV spectra, likely due to scattering of the star's light by circumstellar dust and a potentially larger average dust grain size than typically found in the diffuse interstellar medium (see Massey et al. 2005 and Section 4.2 for further discussion).

There has been very little study of RSGs in the high-energy regime. No RSG has ever been identified as the donor star of a high-mass X-ray binary (a somewhat surprising fact given their large radii and subsequent importance in case B and C mass transfer binaries; see Chapter 5). IR observations *have* identified RSGs as the potential donor stars of some ultraluminous X-ray sources (ULXs; e.g., Heida et al. 2014, 2015), placing them in a close accreting binary with a stellar-mass or even intermediate-mass black hole.

3.2 Effective Temperatures

The effective temperature (T_{eff}) of an RSG is the single most important physical property needed for plotting it on the Hertzsprung–Russell diagram (HRD). By simple definition it dictates a star's position on the HRD x-axis. The cool temperatures of RSGs also give them substantial bolometric corrections; as a result, an accurate luminosity determination for an RSG is strongly dependent on $T_{\rm eff}$.

The minimum $T_{\rm eff}$ of RSGs should correspond to the Hayashi limit, which is the minimum $T_{\rm eff}$ that a star is able to reach while remaining in hydrostatic equilibrium (Hayashi & Hoshi 1961). The Hayashi limit is, in reality, dictated by the very definition of $T_{\rm eff}$, which refers to the temperature at the radius that corresponds to an optical depth (τ) of 1 in a stellar atmosphere. This optical depth is strongly dependent on the opacity, κ, and in cool stars the main source of opacity is H^-, where $\kappa \propto T^9$.

An RSG in hydrostatic equilibrium has a fully convective envelope that is geometrically extended. At the surface of the star $\kappa \propto T^9$; this implies that as the star increases in size, the $T_{\rm eff}$ decreases and the outer layers of the star effectively become transparent. Put another way, as the star's radius increases, in these conditions we must look deeper into the star to reach $\tau = 1$, which happens at an almost constant temperature. This temperature gives us the nearly vertical Hayashi limit on the HRD (this conclusion can also be arrived at analytically by considering the interdependence between luminosity, mass, $T_{\rm eff}$, and opacity for fully convective stars; when H^- is the main source of opacity we wind up with a $T_{\rm eff}$ that is almost completely independent of luminosity and mass).

The Hayashi limit is also metallicity-dependent. The abundance of H^- is proportional to the density of free electrons, which increases with metallicity given that metals are the dominant source of free electrons in stellar atmospheres (they are easier to ionize than H or He). At higher metallicities the H^- opacity therefore increases, leading to a "shallower" optical depth and a cooler Hayashi limit; this effect is seen both theoretically and observationally (see Section 6.4).

Early efforts to determine $T_{\rm eff}$ observationally—and to develop an effective temperature scale mapping $T_{\rm eff}$ to spectral type—were challenging, largely due to the paucity of available data. For lower-mass red giants, interferometric data for nearby stars can be used to assemble an effective temperature scale, but there are too few nearby RSGs to employ this method at higher masses, and simply shifting red giant scales to apply them to RSGs offers only a very rough approximation (e.g., Massey & Olsen 2003). Initial effective temperature scales developed specifically for RSGs used broadband colors to assign $T_{\rm eff}$ based on the small handful of RSGs with measured diameters, but the most commonly used broadband color for RSGs, $(B - V)_0$, is extremely sensitive to surface gravity due to line blanketing effects, making it a poor gauge of $T_{\rm eff}$-dependent shifts in the blackbody continuum (e.g., Lee 1970; White & Wing 1978). There is also a problematic degeneracy between $T_{\rm eff}$-dependent changes in the shape of the continuum and the effects of reddening, a particular problem for mass-losing evolved stars like RSGs that are expected to have significant amounts of circumstellar dust. The early effective temperature scales consistently placed entire RSG populations in the Milky Way and Magellanic Clouds to the right of the Hayashi limit on the HRD, putting them at odds with the predictions of stellar evolutionary theory.

An alternative method for developing an effective temperature scale for RSGs is to make use of the same rich TiO absorption spectrum that dictates their spectral types. Unlike their red dwarf and giant counterparts, the young ages of RSGs mean that, within a given galaxy/population, the abundance of TiO (and the corresponding strengths of the TiO bands) map primarily to $T_{\rm eff}$ rather than long-term variations in surface metal enrichment with time. A cooler $T_{\rm eff}$ corresponds to stronger TiO features (and, by definition, later spectral types).

The new generation of MARCS stellar atmosphere models developed in 2003 was the first to include improved treatments of opacity effects for oxygen-rich molecules, especially TiO (e.g., Gustafsson et al. 1975; Plez et al. 1992; Gustafsson et al. 2003, 2008). Combined with spectrophotometric observations of RSG populations, these models made it possible, for the first time, to use the TiO bands for robust determinations of $T_{\rm eff}$ and the development of an effective temperature scale in RSGs. This was first done in Levesque et al. (2005) for Galactic RSGs, and later extended to RSGs in the Magellanic Clouds (Levesque et al. 2006), using the same temperature-sensitive TiO bands that serve as spectral type diagnostics. The resulting new effective temperature scale was much warmer across RSG spectral subtypes (Figure 3.2), and brought the Milky Way and Magellanic Cloud RSG populations into excellent agreement with the predictions of stellar evolutionary models (although some individual stars do still have a $T_{\rm eff}$ that places them to the right of the Hayashi limit; see Chapters 6 and 7).

It is important to note that use of stellar atmosphere models and optical spectra to determine $T_{\rm eff}$ for RSGs is fundamentally limited by combining a short wavelength range with imperfect treatments of atmospheric geometry. Wavelength-dependent optical depth effects become important in the extended atmospheres of RSGs and are difficult to simulate in one-dimensional models. Put simply, we observe a different stellar "surface" at different wavelengths, and would thus measure a different $T_{\rm eff}$ if this wavelength-dependent effect is not accounted for in atmosphere models (e.g., Davies et al. 2013).

Color-based methods for determining RSG $T_{\rm eff}$ are also in use, and can be used in part to mitigate the effects of working in a single wavelength regime. $(B-V)_0$ colors are, as previously noted, extremely sensitive to surface gravity due to strong line blanketing effects in the B-band (see also Section 3.6), but $(V-K)_0$ colors are a good alternative method for determining $T_{\rm eff}$. Using the relationship between $T_{\rm eff}$, $(V-K)_0$, and the bolometric correction at K based on the MARCS models, calibrations directly relating $(V-K)_0$ colors and $T_{\rm eff}$ in RSGs are now available across a broad range of metallicities and follow the form of $T_{\rm eff} = A - B(V-K)_0 + C(V-K)_0^2 - D(V-K)_0^3$ (see Table 3.1 for the coefficient values at a variety of metallicities). In general, $T_{\rm eff}$ determinations from the molecular band strengths and the $(V-K)_0$ colors agree to within ~100 K, with the $(V-K)_0$ colors generally giving slightly warmer $T_{\rm eff}$ values (e.g., Levesque et al. 2005, 2006).

Overall, the typical range of $T_{\rm eff}$ for Milky Way RSGs is ~4300 K (corresponding to an "early K" spectral type, e.g., K0 or K1) to 3450 K (corresponding to a spectral type of ~M5); however, there are occasional stars with warmer or cooler $T_{\rm eff}$ that fall into the RSG class and this range varies with metallicity (for a detailed discussion see Section 6.4).

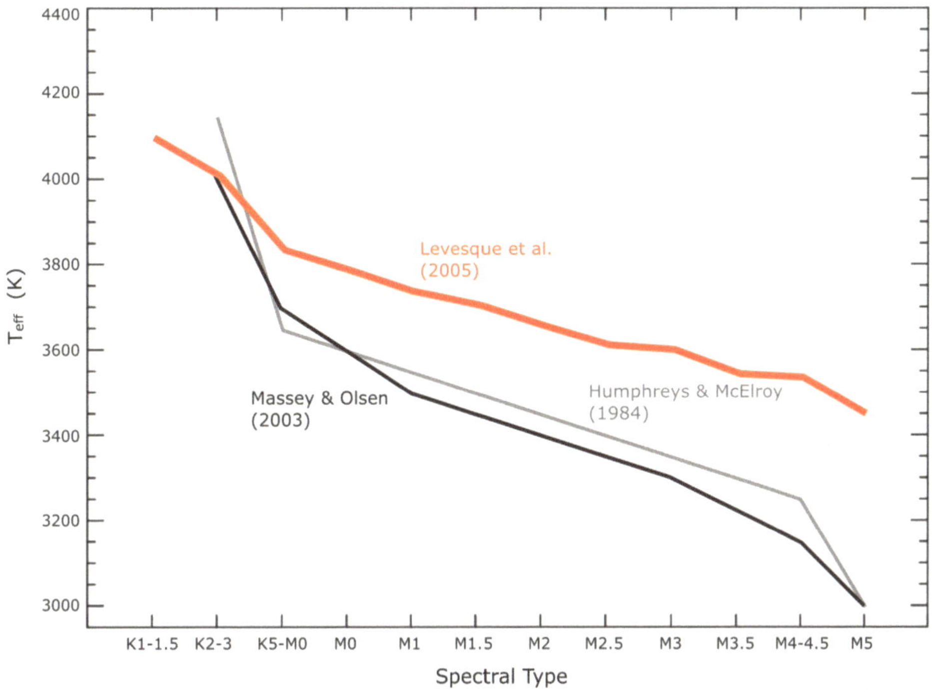

Figure 3.2. Reproduced from Levesque et al. (2005); the effective temperature scale for Galactic RSGs, with T_{eff} determined from the strengths of the TiO bands and fits with the MARCS stellar atmosphere models (red), as compared to previous (and colder) effective temperature scales from Massey & Olsen (2003) and Humphreys & McElroy (1984).

Table 3.1. Coefficients for the T_{eff}-$(V–K)_0$ Relation

[Fe/H]	A	B	C	D	Ref[a]
0.25	8130.9	2113.32	327.883	17.7886	1
0.00	7741.9	1831.83	263.135	13.1943	2
−0.30	7621.1	1737.74	241.762	11.8433	3
−0.50	8280.5	2132.57	318.975	16.9545	…
−0.75	7167.5	1374.20	157.000	6.0481	3
−0.90	8088.0	1968.48	282.832	14.9529	…

Note. [a]References: (1) Massey et al. (2009); (2) Levesque et al. (2005); (3) Levesque et al. (2006).

Finally, it is important to note that all of these methods are complicated by both the photometric and spectroscopic variability of RSGs (see Chapter 7). At the moment, the recommended best means of determining T_{eff} for RSGs is to use contemporaneous observations of V, K, and the star's spectrum wherever possible.

3.3 Masses

When considering the masses of RSGs it is simplest to work in terms of their initial masses—RSGs may shed a significant amount of mass during their lifetimes, but when placing them in a stellar evolution context and discussing their role as supernova progenitors, initial mass is generally the clearest and simplest definition to use.

The theoretical initial mass range of RSGs is limited on the lower end by the minimum initial mass for massive stellar evolution ($\sim 8M_\odot$). While this is a fairly clear-cut theoretical definition, it is a bit more difficult to identify observationally, as in this region of the HRD it becomes difficult to distinguish between low-mass RSGs and "super-AGB" stars (see Chapter 2).

The upper limit of the RSG initial mass range is theoretically dictated by the effects of the atmospheric Eddington limit. The Eddington limit is, descriptively, the point where the force due to radiation pressure is able to overcome gravity somewhere in the star's atmosphere. The atmospheric Eddington limit is given by $L_E^{\rm atm} = 4\pi cGM/\kappa_{\rm atm}$, where $\kappa_{\rm atm}$ is the maximum opacity in the atmosphere across a broad range of optical depths ($10^{-2} < \tau < 10^3$). This opacity peaks around $T_{\rm eff} \sim 10000$ K, leading to a corresponding dip in $L_E^{\rm atm}$. Sufficiently massive stars that are cooling off and evolving rightward on the HRD will then hit this local minimum of the Eddington limit, become unstable, and suffer severe mass loss due to their loosely bound atmospheres (see Ulmer & Fitzpatrick 1998; Lamers & Levesque 2017). This precludes them from continuing to cool off and evolve to the right, stopping stars that hit this limit from becoming RSGs. $L_E^{\rm atm}$ is also inversely proportional to metallicity (due to the metallicity dependence of $\kappa_{\rm atm}$), which should correspond to the formation of RSGs from stars with higher initial masses at lower metallicities.

This effect was originally identified observationally as the Humphreys–Davidson limit (Humphreys & Davidson 1979). A survey of massive stars in the Milky Way and Large Magellanic Cloud revealed a clear absence of luminous cool stars, corresponding to an upper mass limit of $\sim$50–60$M_\odot$; in their paper Humphreys & Davidson noted that this was likely due to the effects of significant mass loss on the evolution of these extremely massive stars. While the exact mass limit is still being refined, this result has since been observationally confirmed for a number of massive star populations in the Local Group, with a clear lack of $\sim$25–40$M_\odot$ RSGs (it is worth noting that there is a further distinction between the mass range of stars that *end* their lives as RSGs and stars that pass through the RSG phase before evolving left again on the HRD to become yellow supergiants or Wolf–Rayet stars—for further discussion see Chapter 8).

3.4 Luminosities

Observationally, luminosity winds up functioning as a proxy for mass when plotting RSGs on the HRD—in practice, the two parameters are degenerate around the Hayashi track region of the HRD, as the evolutionary tracks become nearly vertical during core He fusion (see the discussion in Chapter 2) and the luminosities of stars with different initial masses begin to overlap considerably. Accurate luminosity determinations for RSGs require that their distances and temperatures are also well-known—the former is necessary for applying the correct distance modulus, while the latter is crucial due to the significant bolometric corrections required for RSGs as a result of their cool temperatures. Normal RSGs are also known to be variable in the *V* band by at least a magnitude (see Chapter 7); it is therefore better to use RSGs' *K* band magnitudes to measure luminosity. RSGs are *not* generally variable in *K* (e.g., Josselin et al. 2000; Massey et al. 2009) and the more stable magnitude eliminates the need for contemporaneous observations of T_{eff}, though there is still a significant bolometric correction in *K*. These bolometric corrections are metallicity-dependent and can be calculated using stellar atmosphere models; they follow the form of $BC_K = A - B(T_{\text{eff}}/1000\text{K}) + C(T_{\text{eff}}/1000\text{K})^2$ (see Table 3.2 for the coefficient values at a variety of metallicities).

The predictions of the atmospheric Eddington limit suggest that the maximum RSG mass—and, therefore, the maximum luminosity—should be metallicity-dependent, with lower-metallicity RSGs having higher masses and luminosities. Surprisingly, this has not been borne out by observations of RSGs in the Local Group; Massey et al. (2009) find a consistent maximum luminosity for RSGs of log $(L/L_\odot)$~5.2–5.3 across the ~0.9 dex spread in metallicity between M31 and the Small Magellanic Cloud. However, there are several potential explanations for why this may be an observational effect. RSG populations are typically initially identified based on their broadband colors, and these colors must shift to the blue at decreasing metallicities to accommodate the warmer Hayashi track and corresponding warmer average temperature of the RSG population. When these colors begin to approach the yellow supergiant regime foreground contamination becomes a significant problem in extragalactic samples. The contamination rate is between 50% and 95% for Local Group galaxies (e.g., Drout et al. 2009; Neugent et al. 2012), and while the surface-gravity-sensitive color–color method can be used to distinguish between foreground dwarfs and background supergiants at cooler temperatures (Massey 1998, see Section 3.6), no such method is available for yellow foreground contaminants. This makes it difficult to fully include the warmest members of the RSG population in the sample. It is also possible that mass loss in RSGs is at least partially luminosity-dependent; the most luminous members of the RSG population could therefore be dust-enshrouded, making them difficult to

Table 3.2. Coefficients for the BC_K-T_{eff} Relation

[Fe/H]	A	B	C	Ref[a]
0.25	7.149	1.5924	0.10956	1
0.00	5.574	0.7589	…	2
−0.30	5.502	0.7392	…	3
−0.50	5.403	0.712	…	…
−0.75	5.369	0.7029	…	3
−0.90	5.359	0.699	…	…

Note. [a]References: (1) Massey et al. (2009); (2) Levesque et al. (2005); (3) Levesque et al. (2006).

detect and their luminosities difficult to calculate accurately (see Section 4.3 for further discussion). Finally, the degeneracy between mass and luminosity on the HRD can be complicated, particularly at the upper end, by the effects of rotation, binarity, and mass loss, all of which can diminish the extent to which a star evolves vertically up the Hayashi track during the RSG phase.

The lower luminosity limit for RSGs is, like the lower mass limit, complicated observationally by contamination effects from luminous lower-mass stars at the tip of the red giant branch or asymptotic giant branch. It is also expected to be metallicity-dependent, with a higher lower luminosity limit at lower metallicities. Estimates from stellar evolutionary tracks suggest a conservative lower limit (free of significant contamination from non-massive stars, though there is still some) of log $(L/L_{\odot})$ ~4.0 for Galactic RSGs and $\log(L/L_{\odot})$ ~4.5 for SMC RSGs. While these cutoffs still risk some contamination from "super"-AGB stars, the resulting sample should be free of significant contamination from non-massive stars without significantly excluding the lower-mass end of the true RSG sample.

3.5 Radii

At its simplest, the radius of an RSG can be calculated using the straightforward $L = 4\pi\sigma R^2 T_{\mathrm{eff}}^4$ relation, adopting the radius where T_{eff} is measured as representing the "surface" of the star. In the Milky Way this corresponds to a broad range of radii for RSGs, spanning from ~$100R_{\odot}$ in the smallest cases to maximum radii of as much as ~$1500R_{\odot}$ for normal RSGs (though see Section 5.2 for discussions of unusual or binary RSGs that may potentially have larger effective radii). Like many other physical properties, radius is expected to vary with metallicity, with the dominant T_{eff} dependence ($R \propto T^{-2}$) corresponding to larger radii at higher metallicities due to the cooler temperatures permitted by the Hayashi limit.

Interferometric determinations of radius for RSGs are observationally challenging and limited by the small sample of RSGs with small enough (and sufficiently accurate) distances. Using near-infrared and optical interferometers, angular diameters and corresponding radii have been determined for several dozen red supergiants in recent years, and show generally good agreement with calculations of radii based on luminosity and T_{eff} (e.g., van Belle et al. 2009; Wittkowski et al. 2012, 2017; Arroyo-Torres et al. 2013, 2015). The luminosity- and T_{eff}-based radii estimates for Betelgeuse give an average radius of ~$862R_{\odot}$ (from Levesque et al. 2005 and Smith et al. 2009), while an average of recent interferometric radius estimates for Betelgeuse yields ~$890R_{\odot}$ (Dolan et al. 2016), although the interferometric radius varies with the wavelength of the observations, the definition of the stellar surface, treatments of limb-darkening and asymmetries, and what appear to be actual variations in the physical size of the star (Townes et al. 2009).

As a consequence of their large radii and extended atmospheres, RSGs are generally predicted to have extremely low surface rotation rates, <1 km s^{-1} at the equator (Meynet et al. 2015), although some unusual RSGs such as Betelgeuse have measured surface rotation velocities that are significantly higher, ~15 km s^{-1} (still well below the breakup velocity). It is worth noting, however, that this does not necessarily translate to a decreased angular momentum at deeper layers in the star; this would depend heavily on the degree of differential rotation in the star and to what degree the inner layers of the star are mechanically coupled, which in turn depends on the physical properties of the convective layers and fusion shells in the interior (for more see Chapter 2).

3.6 Surface Gravities

The extremely low surface gravities of RSGs typically range from $\log g = -0.5$ to 0.5. The use of spectral features that are sensitive to surface gravity is a potentially powerful way of distinguishing between RSGs and cool stars with lower luminosity classes (and correspondingly higher surface gravities) such as red dwarfs and giants. This is a particularly important technique in the absence of well-determined distances or when separating background RSGs in nearby galaxies from foreground dwarf and giant populations in the Milky Way (for more discussion see Chapter 6).

Massey (1998) established a means of discriminating between low-gravity RSGs and high-gravity dwarfs photometrically using the $(B - V)$ vs. $(V - R)$ color–color diagram and the effects of surface gravity on the overall shape of the SED. Plotting a sample of red stars in this color space reveals a clear separation in $(B - V)$, with RSGs appearing redder (see Figure 3.3). This is a consequence of the increased

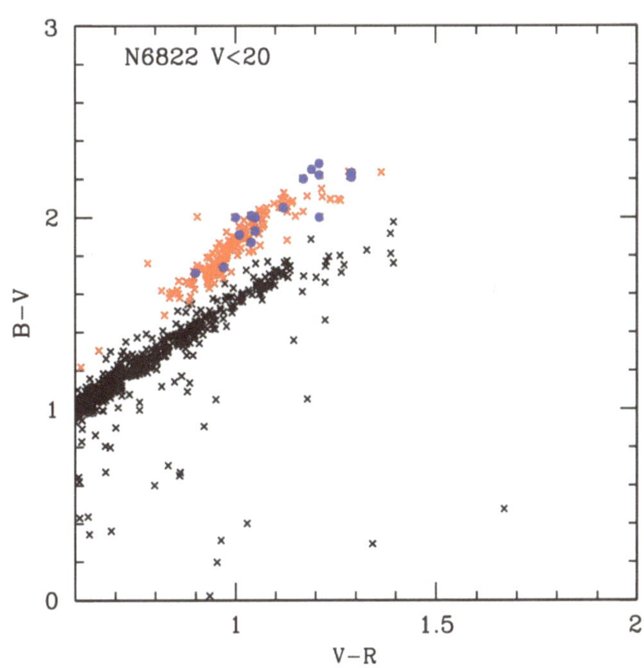

Figure 3.3. Reproduced from Levesque & Massey (2012); red stars in the direction of the Local Group dwarf galaxy NGC 6822 plotted on a $(V - R)$ vs. $(B - V)$ color–color diagram. Following Massey (1998), the RSGs (red) separate from the foreground dwarfs (black) in this color–color space, with the RSGs having notably redder $(B - V)$ colors due to enhanced line-blanketing effects in the B band. Blue points indicate RSGs with spectra obtained in Levesque & Massey (2012).

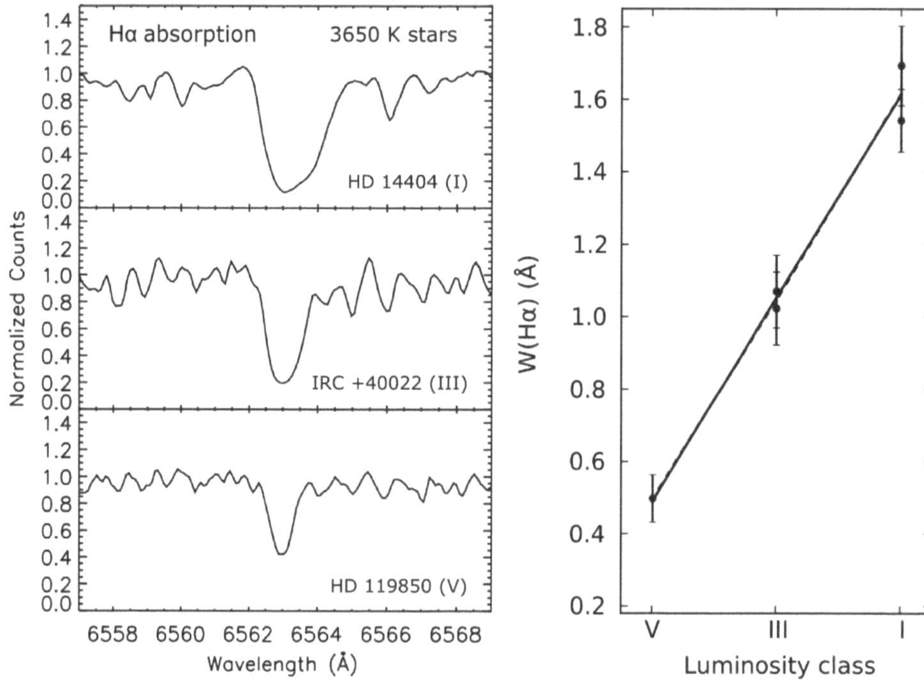

Figure 3.4. Reproduced from Jennings & Levesque (2016). Left: the Hα absorption line in Galactic luminosity class I (top), III (middle), and V (bottom) stars with $T_{\rm eff}$ = 3650 K, showing the progression to narrower lines at higher surface gravities. Right: Hα equivalent width vs. luminosity class for Galactic stars with $T_{\rm eff}$ = 3650 K.

effects of line-blanketing at low surface gravities in these stars, which is particularly influential in the *B* band due to the number of weak metal lines in that regime and serves to decrease the RSGs' *B* fluxes. This color-based method is useful for separating stars at extreme ends of the luminosity class regime but cannot distinguish between supergiants and giants. Surface gravity effects on the SED can also be identified through model atmosphere fitting by evaluating the overall fit to an observed flux-calibrated spectrum. However, this is often hard to quantify due to a degeneracy in the effects of log *g* and the effects of circumstellar reddening on the overall SED shape; it can be done by combining stellar atmosphere models with observed SEDs, but requires iterative fitting to arrive at the best solution. In cases where satisfactory models or well-calibrated SEDs are not available, or where we wish to distinguish between supergiants and giants, we must look for other solutions, such as individual spectral features that are sensitive to surface gravity effects.

The Ca II triplet (8498, 8542, 8662Å) has been shown to be strongly sensitive to surface gravity, with all three lines getting stronger at lower log *g* (e.g., Mallik 1997; Cenarro et al. 2001; Jennings & Levesque 2016). Unfortunately, Ca II is also known to vary with $T_{\rm eff}$ and metallicity, and is blended with the hydrogen Paschen lines at lower spectral resolutions, making it a limited independent gauge of log *g*. The Hα line, while weak in RSG spectra, offers another option that is less sensitive to metallicity and T_{eff}; it is also stronger in supergiants due to the density-dependent overpopulation of the metastable 2s level in hydrogen in non-LTE conditions, with the overpopulation increasing at higher surface gravities when the density is low and radiative and collisional transitions are less frequent (Struve et al. 1939; Huang et al. 2012; Jennings & Levesque 2016; see Figure 3.4).

References

Arroyo-Torres, B., Wittkowski, M., Chiavassa, A., et al. 2015, A&A, **575, 50**

Arroyo-Torres, B., Wittkowski, M., Marcaide, J. M., & Hauschildt, P. H. 2013, A&A, **554, 76**

Bauer, W. H., Bennett, P. D., & Brown, A. 2007, ApJS, **171, 249**

Britavsky, N. E., Bonanos, A. Z., Mehner, A., Garcia-Alvarez, D., Prieto, J. L., & Morrell, N. I. 2014, A&A, **562, 75**

Cenarro, A. J., Cardiel, N., Gorgas, J., et al. 2001, MNRAS, **326, 959**

Davies, B., Figer, D. F., Law, C. J., et al. 2008, ApJ, **676, 1016**

Davies, B., Kudritzki, R.P., Plez, B., et al. 2013, ApJ, **767, 3**

Dolan, M. M., Mathews, G. J., Lam, D. D., et al. 2016, ApJ, **819, 7**

Dorda, R., Negueruela, I., Gonzalez-Fernandez, C., & Tabernero, H. M. 2016, A&A, **592, 16**

Dougherty, S. M., Clark, J. S., Negueruela, I., Johnson, T., & Chapman, J. M. 2010, A&A, **511, 58**

Drout, M. R., Massey, P., Meynet, G., Tokarz, S., & Caldwell, N. 2009, ApJ, **703, 441**

Gazak, J. Z., Davies, B., Kudritzki, R., Bergemann, M., & Plez, B. 2014, ApJ, **788, 58**

Gustafsson, B., Bell, R. A., Eriksson, K., & Nordlund Å. 1975, A&A, **42, 407**

Gustafsson, B., Edvardsson, B., Eriksson, K., et al. 2003, in ASP Conf. Ser. 288, Stellar Atmosphere Modeling, ed. I. Hubeny, D. Mihalas, & K. Werner (San Francisco: ASP), **331**

Gustafsson, B., Edvardsson, B., Eriksson, K., et al. 2008, A&A, **486, 951**

Hayashi, C., & Hoshi, R. 1961, PASJ, **13, 442**

Heida, M., Jonker, P. G., Torres, M. A. P. et al. 2014, MNRAS, **442, 1054**

Heida, M., Torres, M. A. P., Jonker, P.G., et al. 2015, MNRAS, **453, 3510**

Huang, W., Wallerstein, G., & Stone, M. 2012, A&A, **547, A62**

Humphreys, R. M., & Davidson, K. 1979, ApJ, **232, 409**

Humphreys, R. M., & McElroy, D. B. 1984, ApJ, **284, 565**

Jaschek, C., & Jaschek, M. 1990, The Classification of Stars (Cambridge: Cambridge Univ. Press)

Jennings, J., & Levesque, E. M. 2016, ApJ, **821, 131**

Josselin, E., Blommaert, J. A. D. L., Groenewegen, M. A. T., Omont, A., & Li F. L. 2000, A&A, **357, 225**

Lamers, H. J. G. L. M., & Levesque, E. M. 2017, Understanding Stellar Evolution (Bristol: IOP)

Lancon, A., Hauschildt, P. H., Ladjal, D., & Mouhcine, M. 2007, A&A, **468, 205**

Lee, T. A. 1970, ApJ, **162, 217**

Levesque, E. M., & Massey, P. 2012, AJ, **144, 2**

Levesque, E. M., Massey, P., Olsen, K. A. G., et al. 2005, ApJ, **628, 973**

Levesque, E. M., Massey, P., Olsen, K. A. G., et al. 2006, ApJ, **645, 1102**

Mallik, S. V. 1997, A&AS, **124, 359**

Massey, P. 1998, ApJ, **501, 153**

Massey, P., & Olsen, K. A. G. 2003, AJ, **126, 2867**

Massey, P., Plez, B., Levesque, E. M., et al. 2005, ApJ, **634, 1286**

Massey, P., Silva, D. R., Levesque, E. M., Plez, B., & Clayton, G. C. 2009, ApJ, **703, 420**

Menten, K. M., Lundgren, A., Belloche, A., Thorwirth, S., & Reid, M. J. 2008, A&A, **477, 185**
Meynet, G., Chomienne, V., Ekström S., et al. 2015, A&A, **575, 60**
Neugent, K. F., Massey, P., Skiff, B., & Meynet, G. 2012, ApJ, **749, 177**
Plez, B., Brett, J. M., & Nordlund Å. 1992, A&A, **256, 551**
Smith, N., Hinkle, K. H., & Ryde, N. 2009, AJ, **137, 3558**
Struve, O., Wurm, K., & Henyey, L. G. 1939, PNAS, **25, 67**
Townes, C. H., Wishnow, E. H., Hale, D. D. S., & Walp, S. ApJL, **697, L127**
Ulmer, A., & Fitzpatrick, E. L. 1998, ApJ, **504, 200**
van Belle, G. T., Creech-Eakman, M. J., & Hart, A. 2009, MNRAS, **394, 1925**
Verheyen, L., Messineo, M., & Menten, K. M. 2012, A&A, **541, 36**
White, N. M., & Wing, R. F. 1978, ApJ, **222, 209**
Wittkowski, M., Hauschildt, P. H., Arroyo-Torres, B., & Marcaide, J. M. 2012, A&A, **540, 12**
Wittkowski, M., Arroyo-Torres, B., Marcaide, J. M., et al. 2017, A&A, **597, 9**

IOP Astronomy

Astrophysics of Red Supergiants

Emily M. Levesque

Chapter 4

Mass Loss and Dust Production in Red Supergiants

4.1 Winds and Mass-loss Processes

More than half of the mass loss that occurs in a massive star's lifetime happens after it has evolved off the main sequence. A signifiant fraction of this occurs during the RSG phase (Stothers & Chin 1996) when the star's surface gravity is at its lowest. This strong mass loss is, in turn, a critical ingredient in the subsequent evolution of these massive stars. However, the physical properties and mechanisms driving RSG mass loss are still unclear.

Much of our current understanding of RSG mass-loss rates is based on observations of their circumstellar environments. Multi-wavelength observations have revealed evidence of excess circumstellar dust around RSGs, (ΔA_V), which appears to diverge from the standard Cardelli et al. (1989) reddening law with a higher R_V (Massey et al. 2005, 2006; Levesque et al. 2007). There is evidence that this dust is distributed in shells (Danchi et al. 1994) and asymmetrically (e.g., Smith et al. 2001; Schuster et al. 2006; Ohnaka et al. 2008; Scicluna et al. 2015), suggesting that the mass loss driving this dust production is episodic or even eruptive in nature. Levesque et al. (2007) and Massey et al. (2007) identified a sample of unusual RSGs that show extreme photometric and spectroscopic variability on a timescale of months; this variability includes dramatic changes in the observed amount of circumstellar dust and offers further evidence of sporadic mass-loss events (for more discussion see Section 7.2).

A number of different mechanisms have been proposed that could contribute to driving RSG mass loss. The most commonly cited model invokes a stellar wind powered by radiation pressure on dust grains in a similar scenario to that proposed for asymptotic giant branch stars (e.g., van Loon et al. 2006; Höfner 2008). Radiation pressure on *spectral lines*, a dominant driver of stellar winds at higher temperatures, is present in cool stars, but makes only a very small contribution to the

doi:10.1088/978-0-7503-1329-2ch4

radiative force, insufficient for driving a stellar wind on its own (e.g., Bennett 2010). However, the absorption of stellar photons by dust grains in the close circumstellar environment (within several stellar radii of the photosphere) can produce an analogous radiation pressure effect, with the outwardly accelerated dust grains colliding with gas atoms and molecules and dragging them along to produce the stellar wind. At face value this seems to be a very satisfactory and straightforward explanation for mass loss in RSGs, particularly as the collisional coupling of dust to gas in these stars increases as a function of dust grain size, which is expected to be larger in RSGs (see MacGregor & Stencel 1992 and Section 4.2). However, there are several complicating factors that suggest this explanation is incomplete.

As with any mass-loss prescription based on radiation pressure, this model suggests that mass loss ($\dot{M}$) in RSGs should be a function of luminosity, a fact reflected in most existing mass-loss rate prescriptions for RSGs. Mass-loss rates for RSGs can be observationally determined, and most are based on mid-IR emission from their circumstellar dust (with the exception of a handful of RSGs whose circumstellar gas can be directly detected; see further discussion below). Translating these observations into an empirical mass-loss rate prescription was initially phenomenologically agnostic and based on relating the estimated $\dot{M}$ to other physical properties such as L, R, M, and $T_{\rm eff}$.

The first widely used relation was the Reimers law (Reimers 1975; Kudritzki & Reimers 1978), developed for a small sample that included both red giants and red supergiants, where $\dot{M} = 5.5 \times 10^{-13} LR/M$ with L, R, and M in solar units. A decade later, the widely cited de Jager et al. (1988) mass-loss rate relation for RSGs was developed based on observations of 15 RSGs, and is expressed as a sum of Chebyshev polynomials. It has since been used in a number of stellar evolution codes and is often cited as showing good agreement with more recent observations of massive star samples both in and beyond the Milky Way (e.g., Mauron & Josselin 2011; Beasor & Davies 2016).

More recently, van Loon et al. (2005) developed a mass-loss rate relation based on IR spectra of dusty RSGs and oxygen-rich AGB stars in the Large Magellanic Cloud and assuming the dust-driven wind model. Their relation is given by $\log_{10}(\dot{M}) = -5.5 + 1.05 \log_{10}(L/10000 L_\odot) - 6.3 \log_{10}(T_{\rm eff}/3500{\rm K})$. It is important to note that, while this relation is empirically determined from observations, it is based on only a small sample of M-type stars that blends both RSGs and AGB stars. The sample also includes extreme dusty members of the RSG sample (such as the OH/IR star WOH G64—see Section 4.3—and stars with atypically late spectral types for the LMC—see Section 6.4), and may be biased toward larger values of $\dot{M}$ due to the limitations of the mid-IR data available at the time (see Mauron & Josselin 2011).

Unfortunately, it has long been recognized that there is a substantial scatter when plotting L vs. $\dot{M}$ for the observed mass-loss rates of red supergiants (e.g., Jura & Kleinmann 1990; Josselin et al. 2000; Massey et al. 2005; Mauron & Josselin 2011; Meynet et al. 2015; see Figure 4.1), and a number of efforts have focused on resolving or explaining this scatter. Some of the discrepancies can be accounted for by considering the different interpretations employed to translate mid-IR photometry or spectroscopy into mass-loss rates, and these often depend on assumptions

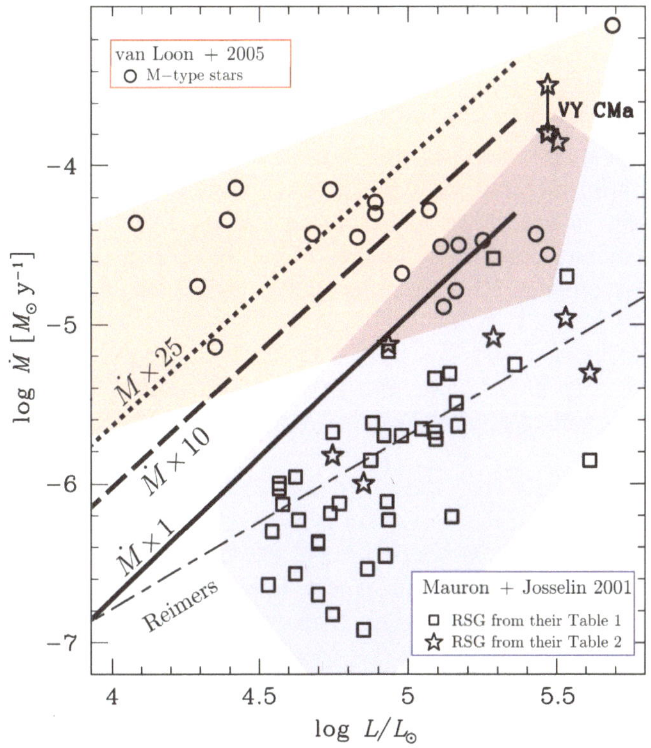

Figure 4.1. Reproduced with permission from Meynet et al. (2015; © ESO); a comparison of RSG mass-loss rates deduced from spectroscopy by van Loon et al. (2005; region shaded in red) and Mauron & Josselin (2011; region shaded in blue). The two stars linked with a vertical line are mass-loss rates for VY CMa from Mauron & Josselin (2011; lower) and Matsuura et al. (2014; upper). The solid, dotted, and dashed lines correspond to averaged RSG mass-loss rates from the Geneva models corresponding to 1, 10, and 25 times the standard mass-loss rate respectively. The short-long dashed line shows the Reimers law for $T_{\rm eff}$ = 3750 K.

of a particular gas-to-dust mass ratio, wind-expansion velocities, distances, and radiative transfer treatments of the dust (e.g., Mauron & Josselin 2011). The L versus $\dot{M}$ parameter space also neglects $T_{\rm eff}$ dependence, though this is small and not enough to completely resolve the scatter. A metallicity dependence may also contribute to this scatter (as these comparisons often include both Milky Way and Magellanic Cloud RSGs and thus span a range of metallicities); Mauron & Josselin (2011) suggest applying a metallicity-scaling factor of $(Z/Z_{\odot}^{0.7})$ to the de Jager et al. (1988) prescription, but once again this is not enough to eliminate the scatter in luminosity. It is also possible that there is an upper limit to the luminosity dependence of RSG mass-loss rates. While a luminosity dependence for $\dot{M}$ is seen observationally (e.g., Bonanos et al. 2010; Beasor & Davies 2016), it appears that it may reach an upper limit and effectively "saturate" at $\log(L/L_{\odot})$ ~5.3 in the LMC (e.g., Bonanos et al. 2010; Mauron & Josselin 2011). All of this is at odds with the straightforward L-dependence that would be expected for a mass-loss rate in which the dominant contribution is from radiation pressure.

The presence of the critical circumstellar dust required to produce a sufficiently strong dust-driven wind is also difficult to test observationally. With the exception of the most nearby RSGs (including Betelgeuse and VY Canis Majoris), circumstellar dust shells have generally been detected at larger distances (tens of stellar radii, e.g., Danchi et al. 1994; Harper et al. 2001; Kervella et al. 2016) where the dust is too far away to be dynamically important in dust-driven mass loss. However, this is largely due to the limits of interferometry and contrast issues with direct imaging of the circumstellar environment (though recent efforts have started successfully obtaining high-contrast imaging of RSG circumstellar environments at closer radii; Scicluna et al. 2015).

Finally, by definition, dust-driven stellar winds also require that some amount of circumstellar dust is *already present* at close radii to the stellar photosphere. This demands another source of mass loss that can produce a sufficient initial quantity of circumstellar dust.

RSGs can potentially lose mass via non-wind mechanisms, such as mass loss induced by radial pulsations. Pulsation is a commonly cited mechanism for aiding the formation of dust-driven winds in cool stars. While the exact mechanism for driving mass loss from pulsations is not well understood, the basic concept is straightforward: the pulsating star effectively "throws" its outer layers to larger radii (possibly through a rebound effect as outer layers "fall" onto deeper ones during a contraction, or simply by decreasing the radius of the stellar photosphere in an atmosphere that is not strongly mechanically coupled). This creates a much smaller density gradient in the outer layers of the star and eventually results in stellar material at large radii—well past the stellar photosphere—that can condense into dust (e.g., Bowen 1988; van Loon et al. 2008; Yoon & Cantiello 2010). There is also observational evidence for the kind of episodic mass loss that would result from these pulsations, including the dust shells observed around RSGs by Danchi et al. (1994) and the rapid variations in $T_{\rm eff}$ and circumstellar dust seen in some of the coldest and most luminous RSGs in the Magellanic Clouds (Levesque et al. 2007; Massey et al. 2007). The large convective cells and high convection speeds in RSG envelopes could also aid mass loss, producing turbulent pressure that can decrease the effective gravity to the point where weaker mechanisms such as line-driven winds could significantly contribute (e.g., Josselin & Plez 2007).

Dissipation of Alfvén waves (e.g., Hartmann & Avrett 1984) is another proposed mechanism that could power mass loss in cool stars. Perturbations in the magnetic field lines of the star will propagate outward as Alfvén waves, which have relatively large damping lengths and make it possible to accelerate a wind by transferring the energy and momentum of the waves to the close circumstellar gas. The degree to which this effect contributes to the total mass loss of RSGs remains unclear. The original Hartmann & Avrett (1984) model was developed to simulate Alfvén wave-driven mass loss in Betelgeuse, but required electron temperatures in the gas that were notably higher than those predicted by later empirical models (e.g., Harper et al. 2001). Alfvén waves only couple to ionized gas, so for the mostly neutral circumstellar gas expected around very cool stars they are not expected to contribute significantly to mass loss. However, it is worth noting that Betelgeuse is a relatively cool RSG; for early-to-mid-K-type RSGs with warmer effective temperatures, higher surface gravities, and more compact stellar atmospheres and circumstellar environments, it is possible that the Alfvén wave scenario could play a more substantial role (e.g., Charbonneau & MacGregor 1995; Vidotto & Jatenco-Pereira 2006). This source of mass loss also requires a magnetic field of several Gauss, but this has been seen in at least some RSGs. Auriére et al. (2010) and Tessore et al. (2017) observed magnetic fields of ~1 G in Betelgeuse, while Vlemmings et al. (2005) observed circular polarization from Zeeman-splitting in the H_2O masers surrounding VX Sgr (see below), suggesting a surface magnetic field as strong as 4 G. Most recently, Tessore et al. (2017) detected Betelgeuse-like ~1 G magnetic fields in the RSGs CE Tau and μ Cep, and a magnetic field with a strength of several G in the RSG α Her. It is possible that strong magnetic fields in RSGs could be generated by a dynamo action driven by differential rotation between the expanding outer envelope and the contracting core (Uchida & Bappu 1982; see Chapter 2).

Another phenomenon dubbed "supra-Eddington mass loss" is expected to occur in RSGs. The same opacity-dependence in the Eddington limit that produces the Humphreys–Davidson limit (see Chapter 3) can also produce critical instabilities within RSGs. Variations in the ionization level of hydrogen beneath the surface of the star could lead to a scenario in which the Eddington luminosity is decreased in some layers, possibly dropping below the actual stellar luminosity. The precise effect of this is difficult to model, as the star would no longer be in hydrostatic equilibrium (Maeder 2009), but this is expected to lead to significant mass loss. The Geneva evolutionary models simulate this by artificially increasing the "normal" expected RSG mass-loss rate by a factor of 3 during the supra-Eddington phase and find that the overall time-averaged mass-loss rates of their model RSGs agree well with the van Loon et al. (2005) mass-loss rates (e.g., Ekström et al. 2012; Meynet et al. 2015).

It is clear that fully explaining and quantifying mass loss in RSGs, and how $\dot{M}$ relates to other physical parameters, is a difficult proposition. At the same time, it is critically important for understanding RSGs and the larger role that they play in massive star evolution. Meynet et al. (2015) demonstrated that even modest modifications to the RSG mass-loss rate can result in dramatically different theoretical evolutionary tracks (see Figure 4.2). This powerfully illustrates the importance of mass loss during the RSG phase to the terminal physical properties of the massive star, and to the evolution of the star's circumstellar environment.

4.2 Dust from Supergiants

As noted above, circumstellar dust, formed as a consequence of grain condensation during stellar mass loss, is the best observable means of studying mass loss in RSGs (and may also be one of its drivers). Most RSGs are known to have at least some amount of circumstellar dust, and its geometry, distribution, and composition is key for understanding RSG mass loss and its contribution to the interstellar medium.

The first detections of circumstellar dust shells around RSGs came in the late 1960s. Near-infrared photometry and spectroscopy suggested the presence of extensive circumstellar clouds or shells around the extreme RSGs NML Cyg and

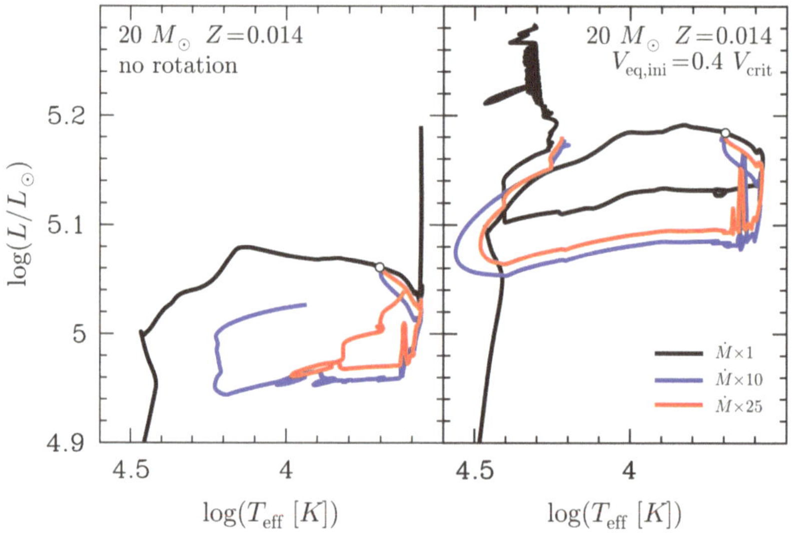

Figure 4.2. Reproduced with permission from Meynet et al. (2015; © ESO); partial evolutionary tracks for $20M_\odot$ non-rotating (left) and rotating (right) stellar models, simulating a standard mass-loss rate (black) along with mass-loss rates enhanced by factors of 10 (blue) and 25 (red). The beginning of the portion of the track computed with an enhanced RSG mass-loss rate is marked with a small white circle.

VY CMa, based on observations that suggested a very high luminosity and a large infrared excess (Johnson 1968; Hyland et al. 1969). Later observations revealed that circumstellar dust shells are not unique to these extreme RSGs, but are instead common in the Galactic RSG population as a whole (Stencel et al. 1988, 1989). The dust production rate and the excess reddening imposed on the star by this dust both correlate with the stellar luminosity (Massey et al. 2005; Verhoelst et al. 2009).

The current best model for dust production and condensation in RSGs adopts the sequence described in Tielens (1990) for oxygen-rich circumstellar regions, with the nucleation of Al_2O_3 (alumina) at ~1760 K to form the silicate $Ca_2Al_2SiO_7$ (augite) and later $(Mg,Fe)_2SiO_4$ (olivines); RSGs appear to follow a different dust evolution sequence than AGB stars (see Speck et al. 2000). The mass fraction of these silicates also appears to correlate with mass-loss rate, in agreement with the proposed condensation sequence. Infrared spectra also suggest the presence of an additional dust species, most likely amorphous carbon given the detection of emission bands from polycylcic aromatic hydrocarbons (PAHs) in some RSG spectra (Verhoelst et al. 2009), distinguishing RSG dust from that produced by AGB stars.

The extinction curve for this circumstellar dust also appears to differ from that of normal interstellar dust. Most of the extinction caused by RSG circumstellar dust will occur in a thin shell close to the star, as this is where the dust first condenses and where the mass density will be the highest (e.g., Massey et al. 2005). Early observations of NML Cyg and VY CMa found a larger relative extinction in the infrared compared to the optical, suggesting a larger-than-average grain size. Scicluna et al. (2015) estimated a dust grain size of 0.5μm—50 times larger than that of the diffuse ISM—around VY CMa. Similar results were found based on observations of the circumstellar envelope surrounding the α Sco binary system (comprised of an RSG primary and a main-sequence B-type secondary; see Section 5.2), suggesting that the dust grains of the envelope were larger and siliceous rather than carbonaceous (Snow et al. 1987). The Levesque et al. (2005) survey of Galactic RSGs also noted that many RSG spectra had excess flux in the near-ultraviolet when compared to the predictions of stellar models (see Figure 4.3). Closer examination of this excess UV flux found it to be closely correlated with the amount of circumstellar reddening, which is potentially explained by larger dust grain sizes that produce less UV extinction or by scattering of UV light into the observer "beam" by off-axis portions of the unresolved circumstellar dust shell (Massey et al. 2005). Multi-wavelength observations of μ Cep concluded that its circumstellar dust did not follow a standard reddening law, with a much larger R_V than the standard value of 3.1 adopted by the Cardelli et al. (1989) reddening law for the diffuse Galactic interstellar medium (see Levesque 2010 for a discussion).

The unusual nature of the circumstellar dust produced by RSGs has interesting implications both for studying dust-enshrouded RSGs (see Section 4.3) and for observing and modeling the interstellar medium in young massive star clusters or high-redshift galaxies. In such environments, where significant populations of lower-mass cool stars have not yet evolved, RSGs are expected to contribute significantly

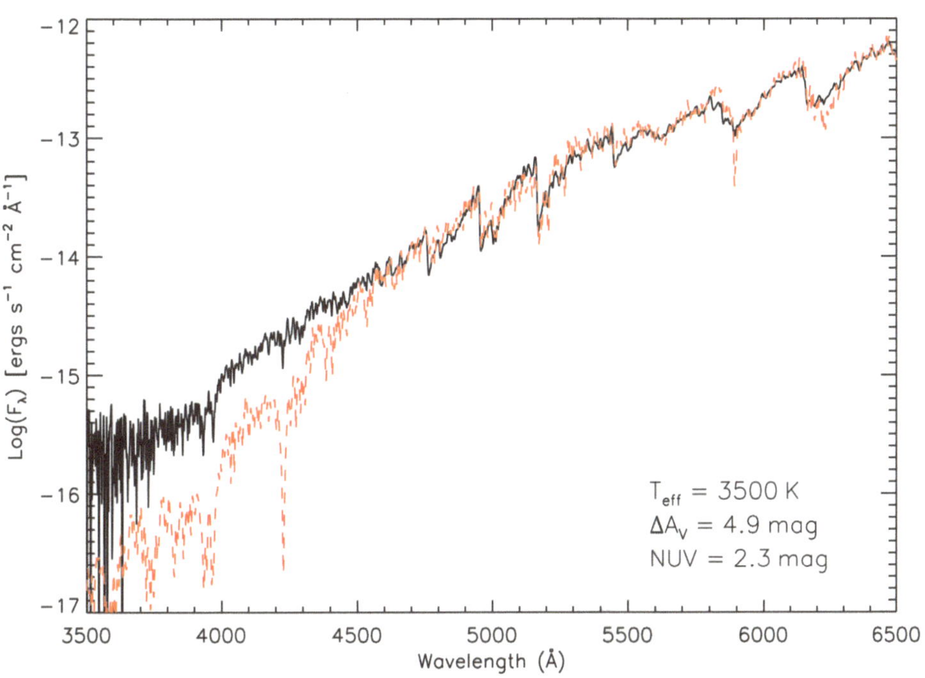

Figure 4.3. Reproduced from Massey et al. (2005); the observed spectral energy distribution of the Galactic RSG KY Cyg (solid black) compared to the best-fit MARCS stellar atmosphere model (dashed red); the model has been "reddened" to correct for extinction. KY Cyg shows a considerable amount of excess NUV flux (~2.3 mag) compared to the theoretical atmosphere model. KY Cyg also shows a considerable amount of excess reddening, $\Delta A_V = 4.9$ mag, compared to the A_V of its host association, Cyg OB1.

to dust production—they will be the dominant dust producers in metal-poor environments where evolved dust-producing Wolf–Rayet stars are expected to be relatively rare (see Chapter 6). The large grain size and atypical composition of RSG dust could therefore significantly skew R_V in these environments. Circumstellar RSG dust is even expected to potentially survive sublimation by the star's eventual supernova (e.g., Scicluna et al. 2015), though it is also possible that the newly formed dust observed in the close circumstellar environment could be gradually broken down into smaller grain sizes before mixing with the interstellar medium (e.g., Jura 1996; Whittet 2003; Massey et al. 2005). If these stars are indeed contributing dust to the ISM that does not follow the standard Cardelli reddening law, this could dramatically impact how we interpret observations of young star-forming galaxies and correct for their *in situ* dust effects.

4.3 Extreme Mass Loss in RSGs - OH/IR Supergiants

A small number of RSGs in the Milky Way and Magellanic Clouds are set apart from the general population by a unique set of physical properties attributable to mass-loss behavior. These stars show evidence of thick asymmetric circumstellar envelopes—marking them as extremely bright IR sources—and circumstellar SiO, H_2O, and OH maser emission. Their strong OH maser activity and high IR luminosities lend them their designation as "OH/IR" supergiants.

Maser activity in RSGs, and its association with mass-loss behavior, is actually not uncommon. SiO, H_2O, and OH masers are all produced in the outflowing material generated by episodes of high mass loss and are detected in the circumstellar environments of both RSGs and asymptotic giant branch stars (it is also worth noting that the giants' circumstellar masers are narrower than those produced by supergiants; Verheyen et al. 2012). SiO masers are the most commonly detected masers in RSG atmospheres; they are formed low in the stellar winds where the outflow velocity is low and have radial velocities very similar to the star's systemic velocity (Jewell et al. 1991). At larger radii, SiO is depleted by silicate dust grain formation and H_2O masers are formed, while at the largest distances from the star H_2O is photodissociated to OH (see Habing 1996). Recent studies have revealed that $\sim$15% of Galactic RSGs show associated maser activity (Verheyen et al. 2012). It is possible that maser emission is a phase of RSG evolution that either ignites or intensifies at high mass-loss rates or at late phases of the stars' evolution up the Hayashi track (Davies et al. 2008).

That said, detecting the suite of all three circumstellar masers combined with a thick dust nebula is sufficient to set OH/IR stars apart as distinct from "normal" RSGs. These circumstellar nebulae can also make it especially difficult to determine the physical properties of the central star. Multiple scatterings of the photons in the dust shell can dilute or even fully obscure the absorption lines in the stellar spectrum, known as "veiling" (e.g., Romanik & Leung 1981). One extreme example of this veiling effect is the M31 RSG J004047.84+405602.6, which is photometrically similar to other M31 RSGs but shows a completely featureless and very red spectrum (Massey et al. 2009). This phenomenon could in turn lead to incorrect estimates of a dust-enshrouded RSG's spectral type or $T_{\rm eff}$, as the crucial TiO bands (see Section 3.2) might be "filled in" or even erased entirely. Failing to properly account for the presence of a dust envelope can also lead to errors in estimating an RSG's bolometric luminosity, which in turn can impact its placement on the Hertzsprung–Russell diagram and even its assumed mass-loss rate. Two famous examples of this are the Milky Way RSG VY Canis Majoris (VY CMa) and the LMC RSG WOH G64.

The circumstellar nebula of VY CMa was first observed in 1917 (Perrine 1923); it was originally identified as a multiple system (See 1897), but later observations determined that objects thought to be stellar companions were instead structures within the nebula. The nebula itself has since been extensively studied; it is asymmetric and highly structured (e.g., Smith et al. 2001, 2004; Humphreys et al. 2005, 2007; Scicluna et al. 2015; see Figure 4.4), and is one of the brightest 5-20μm obejcts in the sky due to heating by the central star (Herbig 1970a). VY CMa also shows strong SiO, H_2O, and OH maser emission (e.g., Knowles et al. 1969; Buhl et al. 1974) which has been studied extensively and used to derive a precise distance measurement to VY CMa of $1.14^{+0.11}_{-0.09}$ kpc (Choi et al. 2008). The presumed mass-loss rate of VY CMa is also the highest of any Milky Way RSG, at ~2.8-6 × 10^{-4} $M_\odot$ yr^{-1} (e.g., De Beck et al. 2010; Shenoy et al. 2016). VY CMa is also extremely variable, with changes on the order of ±3.5 mag extended back over 200 years (Wallerstein 1958; Robinson 1970, 1971; Wallerstein & Gonzalez 2001). It has maintained a spectral type of M3-5 I for over 50 years but displays emission features from neutral atomic lines that vary on a timescale of months, as well as weak TiO bandhead

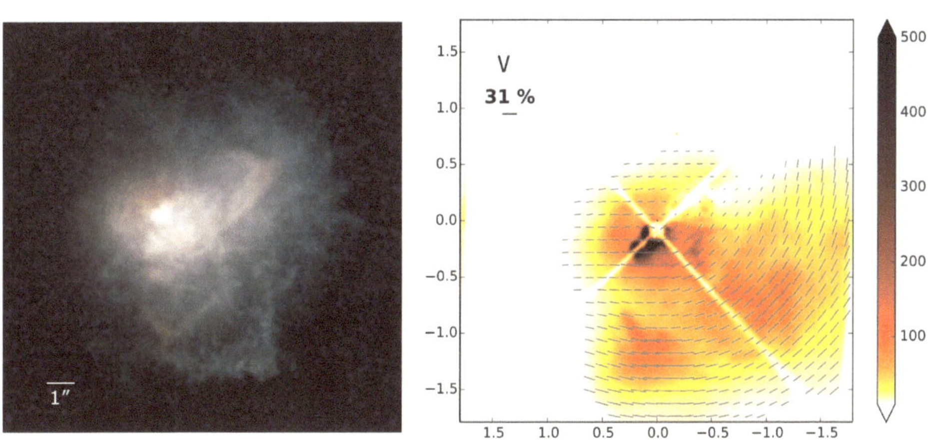

Figure 4.4. Left: reproduced from Smith et al. (2001); composite color *HST* image of VY CMa, including the F410M filter (blue), F547M filter (green), and a combination of the F656N and F1042M filters (red). Right: reproduced with permission from Scicluna et al. (2015; © ESO); polarized intensity in *V* of VY CMa, taken with the Spectro-Polarimetric High-contrast Exoplanet Research (SPHERE) instrument with the Zurich Imaging Polarimeter (ZIMPOL) at the Very Large Telescope. The polarization direction and fraction are shown in the overlaid vector field, with offsets from the central star shown in arcseconds and the color scale in arbitrary units. The cross-shaped structure is produced by the suspended coronagraphic mask obscuring part of the image plane.

emission and CO rotational emission lines (e.g., Wallerstein 1958; Herbig 1970b; Hyland et al. 1969; Wallerstein & Gonzalez 2001; Massey et al. 2006; Matsuura et al. 2014). The TiO band strengths correspond to a $T_{\rm eff}$ of ~3650 ± 25 K (Massey et al. 2006); its bolometric luminosity is more difficult to determine, as this requires accounting for both absorption and re-radiated thermal emission effects from the circumstellar dust. Massey et al. (2006) find an $M_{\rm bol} \sim -7.0$, in good agreement with the predictions of the evolutionary tracks, while Choi et al. (2008) estimate $M_{\rm bol} \sim -9.0$ based on their distance and integrating over the optical and IR fluxes of the SED.

Like VY CMa, WOH G64 in the LMC is extremely bright in the infrared and surrounded by a dusty nebula (e.g., Westerlund et al. 1981; Elias et al. 1986; Roche et al. 1993; Ohnaka et al. 2008). The nebula itself is an asymmetric dusty torus, viewed almost head-on (Ohnaka et al. 2008). WOH G64 is also a SiO, H_2O, and OH maser source; the 86.24335 GHz ($v = 1$, $J = 2 - 1$) SiO maser around WOH G64 was the first extragalactic SiO maser ever discovered (Wood et al. 1986; van Loon et al. 1996, 1998, 2001; Marshall et al. 2004). The maser emission suggests a considerable mass outflow and the presence of two expanding dust shells; its mass-loss rate is estimated to be ~7.6-12 $\times 10^{-4}$ $M_\odot$ $\rm yr^{-1}$ (van Loon et al. 2005; Groenewegen et al. 2009). The strong TiO bands in the WOH G64 spectrum correspond to a very cool $T_{\rm eff}$ of ~3400 ± 25 K (Levesque et al. 2009); combined with the geometrical model for the nebula and its expected contribution to near-IR emission, this leads to an estimated $M_{\rm bol} \sim -8.9$ (Ohnaka et al. 2008; Levesque et al. 2009), in good agreement with predictions of stellar evolutionary tracks but still unusually cold and luminous for an RSG in the LMC. Most interestingly, the spectrum of WOH G64 also has a nebular emission line spectrum that includes Hβ, [OIII] λ5007, [N I] $\lambda\lambda$ 5198,5200, [O I] λ6300, Hα, [N II] $\lambda\lambda$6548,6584, and [S II] $\lambda\lambda$6717,6731, as well as weak TiO bandhead emission and CO rotational emission lines like VY CMa (Elias et al. 1986; Levesque et al. 2009; Matsuura et al. 2016). The excitation mechanism for these lines is unclear. It is possible that the lines could be excited in the circumstellar nebula through ionization by a hot companion, as is seen in VV Cep and other interacting RSG spectroscopic binaries (see Chapter 5); however, the velocities of the nebular emission lines don't support a binary model. The lines could also arise from excitation by a nearby hot star (a scenario that was proposed to explain the strange ionized nebula surrounding the Galactic RSG W26; Wright et al. 2014), or by collisional excitation via shock heating (potentially explaining the unusual line strengths and ratios, which differ considerably from those seen in more typical excitation scenarios such as HII regions).

The small sample of other OH/IR supergiants—including VX Sgr, S Per, and NML Cyg—have all shown evidence of asymmetric dust nebulae (Schuster et al. 2006). With the exception of WOH G64, all of these stars are located in the Milky Way, and their circumstellar masers make it possible to measure their distances extremely accurately. Their proximity, combined with their extended nebulae, have made them excellent observational targets for studying RSG mass loss in detail. They are among the only RSGs for which their winds have been detected through observations of circumstellar *gas*, rather than dust (see Mauron & Josselin 2011).

VY CMa in particular is a popular target for observations with the Atacama Large Millimeter/submillimeter Array (ALMA), using molecular emission to study the inner wind morphology, the dust condensation process, and anisotropic mass loss (e.g., O'Gorman et al. 2015; De Beck et al. 2015; Decin et al. 2016). However, it is important to note that these stars' extreme natures make it hard to extrapolate studies of these phenomena to the larger RSG population. It is still unclear whether all RSGs will pass through an OH/IR supergiant phase, or whether this behavior is relegated to only the coldest, most massive, most luminous, or otherwise atypical cold supergiants.

References

Auriére, M., Donati, J-F., Konstantinova-Antova, R., Perrin, G., Petit, P., & Roudier, T. 2010, A&A, **516, 2**

Beasor, E. R., & Davies, B. 2016, MNRAS, **463, 1269**

Bennett, P. D. 2010, in ASP Conf. Ser. 425, Hot and Cool: Bridging Gaps in Massive Star Evolution, ed. C. Lietherer, P. Bennett, P. Morris, & J. van Loon (San Francisco, CA: ASP), 181

Bonanos, A. Z., Lennon, D. J., Köhlinger, F., et al. 2010, AJ, **140, 416**

Bowen, G. H. 1988, ApJ, **329, 299**

Buhl, D., Snyder, L. E., Lovas, F. J., & Johnson, D. R. 1974, ApJ, **192, 97**

Cardelli, J. A., Clayton, G. C., & Mathis, J. S. 1989, ApJ, **345, 245**

Charbonneau, P., & MacGregor, K. B. 1995, ApJ, **454, 901**

Choi, Y. K., Hirota, Y., Honma, M., et al. 2008, PASJ, **60, 1007**

Danchi, W. C., Bester, M., Degiacomi, C. G., Greenhill, L. J., & Townes, C. H. 1994, AJ, **107, 1469**

Davies, B., Figer, D. F., Law, C. J., Kudritzki, R. P., Najarro, F., Herrero, A., & MacKenty, J. W. 2008, ApJ, **676, 1016**

De Beck, E., Decin, L., de Koter, A., Justtanont, K., Verhoelst, T., Kemper, F., & Menten, K. M. 2010, A&A, **523, 18**

De Beck, E., Vlemmings, W., Muller, S. et al. 2015, A&A, **580, 36**

Decin, L., Richards, A. M. S., Millar, T. J., et al. 2016, A&A, **582, 76**

de Jager, C., Nieuwenhuijzen, H., & van der Hucht, K. A. 1988, A&AS, **72, 259**

Ekström, S., Georgy, C., Eggenberger, P., et al. 2012, A&A, **537, 146**

Elias, J. H., Frogel, J. A., Schwering, P. B. W. 1986, ApJ, **302, 675**

Groenewegen, M. A. T., Sloan, G. C., Soszyński, I., Petersen, E. A. 2009, A&A, **506, 1277**

Habing, H. J. 1996, A&ARv, **7, 97**

Harper, G. M., Brown, A., & Lim, J. 2001, ApJ, **551, 1073**

Hartmann, L., & Avrett, E. H. 1984, ApJ, **284, 238**

Herbig, G. H. 1970a, ApJ, **162, 557**

Herbig, G. H. 1970b, MSRSL, **19, 13**

Höfner, S. 2008, A&A, **491, 1**

Humphreys, R. M., Davidson, K., Ruch, G., & Wallerstein, G. 2005, AJ, **129, 492**

Humphreys, R. M., Helton, L. A., & Jones, T. J. 2007, AJ, **133, 2716**

Hyland, A. R., Becklin, E. E., Neugebauer, G., & Wallerstein, G. 1969, ApJ, **158, 619**

Jewell, P. R., Snyder, L. E., Walmsley, C M., et al. 1991, A&A, **242, 211**

Johnson, H. L. 1968, ApJ, **154, 125**

Josselin, E., Blommaert, J. A. D. L., Groenewegen, M. A. T., Omont, A., & Li, F. L. 2000, A&A, **357, 225**

Josselin, E., & Plez, B. 2007, A&A, **469, 671**

Jura, M. 1996, ApJ, **472, 806**

Jura, M., & Kleinmann, S. G. 1990, ApJS, **73, 769**

Kervella, P., Lagadec, E., Montarges, M., et al. 2016, A&A, **585, 28**

Knowles, S. H., Mayer, C. H., Cheung, A. C., Rank, D. M., & Townes, C. H. 1969, Sci, **163, 1055**

Kudritzki, R-P., & Reimers, D. 1978, A&A, **70, 227**

Levesque, E. M. 2010, NewAR, **54, 1**

Levesque, E. M., Massey, P., Olsen, K. A. G., et al. 2005, ApJ, **628, 973**

Levesque, E. M., Massey, P., Olsen, K. A. G., & Plez, B. 2007, ApJ, **667, 202**

Levesque, E. M., Massey, P., Plez, B., & Olsen, K. A. G. 2009, AJ, **137, 4744**

MacGregor, K. B., & Stencel, R. E. 1992, ApJ, **397, 644**

Maeder, A. 2009, Physics, Formation and Evolution of Rotating Stars (Heidelberg: Springer)

Marshall, J. R., van Loon, J. Th., Matsuura, M. 2004, MNRAS, **355, 1348**

Massey, P., Levesque, E. M., Olsen, K. A. G., Plez, B., et al. 2007, ApJ, **660, 301**

Massey, P., Levesque, E. M., Plez, B. 2006, ApJ, **646, 1203**

Massey, P., Plez, B., Levesque, E. M.,Olsen, K. A. G., Clayton, G. C., Josselin, E. 2005, ApJ, **634, 1286**

Massey, P., Silva, D. R., Levesque, E. M., Plez, B., Clayton, G. C. 2009, ApJ, **703, 420**

Matsuura, M., Sargent, B., Swinyard, B., et al. 2016, MNRAS, **462, 2995**

Matsuura, M., Yates, J A., Barlow, M. J., et al. 2014, MNRAS, **437, 532**

Mauron, N., Josselin, E. 2011, A&A, **526, 156**

Meynet, G., Chomienne, V., Ekström, S., et al. 2015, A&A, **575, 60**

O'Gorman, E., Vlemmings, W., Richards, A. M. S., et al. 2015, A&A, **573, 1**

Ohnaka, K., Driebe, T., Hoffman, K-H., Weigelt, G., & Wittkowski, M. 2008, A&A, **484, 371**

Perrine, C. D. 1923, PASP, **35, 229**

Reimers, D. 1975, MSRSL, **8, 369**

Robinson, L. J. 1970, IBVS, **465, 1**

Robinson, L. J. 1971, IBVS, **599, 1**

Roche, P. F., Aitken, D. K., & Smith, C. H. 1993, MNRAS, **262, 301**

Romanik, C. J., & Leung, C. M. 1981, ApJ, **246, 935**

Schuster, M. T., Humphreys, R. M., & Marengo, M. 2006, AJ, **131, 603**

Scicluna, P., Siebenmorgen, R., Wesson, R., Blommaert, J. A. D. L., Kasper, M., Voshchinnikov, N. V., & Wolf, S. 2015, A&A, **584, 10**

See, T. J. J. 1897, AJ, **18, 181**

Shenoy, D., Humphreys, R. M., Jones, T. J., et al. 2016, AJ, **151, 51**

Smith, N. 2004, MNRAS, **349, L31**

Smith, N., Humphreys, R. M., Davidson, K., Gehrz, R. D., Schuster, M. T., & Krautter, J. 2001, AJ, **121, 1111**

Snow, T. P., Buss, R. H., Gilra, D. P., & Swings, J. P. 1987, ApJ, **321, 921**

Speck, A. K., Barlow, M. J., Sylvester, R. J., & Hofmeister, A. M. 2000, A&AS, **146, 437**

Stencel, R. E., Pesce, J. E., & Bauer, W. H. 1988, AJ, **95, 141**

Stencel, R. E., Pesce, J. E., & Bauer, W. H. 1989, AJ, **97, 1120**

Stothers, R. B., & Chin, C-W. 1996, ApJ, **468, 842**

Tessore, B., Lébre, A., Morin, J., et al. 2017, A&A, **603, 129**

Tielens, A. G. G. M. 1990, in Proc. of the International Colloquium, From Miras to Planetary Nebulae: Which Path for Stellar Evolution? (Gif-sur-Yvette: Éditions, Frontières), 186
Uchida, Y., & Bappu, M. K. V. 1982, ApA, **3, 277**
van Loon, J. Th., Cioni, M-R. L., Zijlstra, A. A., & Loup, C. 2005, A&A, **438, 273**
van Loon, J. Th., Cohen, M., Oliveira, J. M., et al. 2008, A&A, **487, 1055**
van Loon, J. Th., Marshall, J. R., Cohen, M., et al. 2006, A&A, **447, 971**
van Loon, J. Th., te Lintel Hekkert, P., Bujarrabal, V., et al. 1998, A&A, **337, 141**
van Loon, J. Th., Zijlstra, A. A., Bujarrabal, V., & Nyman, L-Å. 1996, A&A, **306, L29**
van Loon, J. Th., Zijlstra, A. A., Bujarrabal, V., & Nyman, L-Å. 2001, A&A, **368, 950**
Verheyen, L., Messineo, M., & Menten, K. M. 2012, A&A, **541, 36**
Verhoelst, T., van der Zypen, N., Hony, S., Decin, K., Cami, J., & Eriksson, K. 2009, A&A, **498, 127**
Vidotto, A. A., & Jatenco-Pereira, V. 2006, ApJ, **639, 416**
Vlemmings, W. H. T., van Langevelde, H. J., & Diamond, P. J. 2005, A&A, **434, 1029**
Wallerstein, G. 1958, PASP, **70, 479**
Wallerstein, G. & Gonzalez, G. 2001, PASP, **113, 954**
Westerlund, B. E., Olander, N., & Hedin, B. 1981, A&AS, **43, 267**
Whittet, D. C. B. 2003, Dust in the Galactic Environment (Bristol: IOP)
Wood, P. R., Bessell, M. S., & Whiteoak, J. B. 1986, ApJ, **306, L81**
Wright, N. J., Wesson, R., Drew, J. E., et al. 2014, MNRAS, **437, 1**
Yoon, S-C., & Cantiello, M. 2010, ApJL, **717, 62**

Emily M. Levesque

Chapter 5

Red Supergiants in Binaries

5.1 Modeling RSG Binaries

The interaction of a massive star with a binary companion through mass transfer can impact its surface abundances, rotation rate, evolutionary pathway, mass loss, and even the terminal core-collapse product. It is therefore extremely important to understand the fraction of massive stars that are in interacting binaries, and the point in their evolution where this interaction occurs.

We already observe a significant binary fraction among main-sequence massive stars: at least 35% of all O stars show some observational signature of a nearby binary companion, and ~26% of O stars are expected to interact with a binary companion during their time on the main sequence (case A mass transfer; e.g., Garmany et a. 1980; Sana et al. 2012), filling their Roche lobe and enabling mass transfer through the first Lagrangian point. It is reasonable to expect that the total interacting binary fraction of massive stars will increase further once mass transfer at the Hayashi limit (case B) is taken into account. The degree to which this increases the total interacting binary fraction of massive stars clearly depends strongly on a careful treatment of RSGs. The mass range of stars that become RSGs directly impacts the percentage of massive stars in a given population that will ever reach the Hayashi limit, and the orbital separations of massive binaries that will eventually experience case B mass transfer directly depend on the maximum radii achieved by RSGs.

Most current work on the RSG phase and its role in interacting binaries is being done theoretically. There are no clear observational examples of RSGs in *true* mass-transfer binaries; however, this is almost certainly a consequence of observational limitations rather than a true lack of such systems. Low-mass or dust-enshrouded companions can be difficult to detect through traditional photometric means. While hot companions can be detected spectroscopically, cool companions pose more of a challenge when trying to deconvolve two K-type or M-type spectra. Most of the mass transfer in the case B scenario is also expected to occur during an extreme short

doi:10.1088/978-0-7503-1329-2ch5

period of time (~$10^3 - 10^4$ years), greatly limiting the number of systems in a population that we can expect to catch in this stage (e.g., Eldridge & Stanway 2016; Lamers & Levesque 2017). Finally, the signature of true mass transfer in cool massive stars is not well-constrained. The observational distinction between companions that are accreting material from the RSG wind (see Section 5.2), transferring mass from the outer layers of the RSG itself, or even undergoing common envelope evolution is (literally) nebulous.

Despite a lack of systems that can be used as observational examples, progress has been made in modeling the overall effects of binary interaction on massive star evolution. Post-main-sequence mass transfer has a very similar effect to extreme mass loss in these stars: a massive star will spend a shorter fraction of its life as an RSG before its outermost layers are stripped off. This results in the star evolving back to the blue side of the H-R diagram and becoming a yellow supergiant or Wolf–Rayet star. The exact onset of this mass transfer and its subsequent effect on the star's evolution depends strongly on the orbital separation of the binary (Figure 5.1).

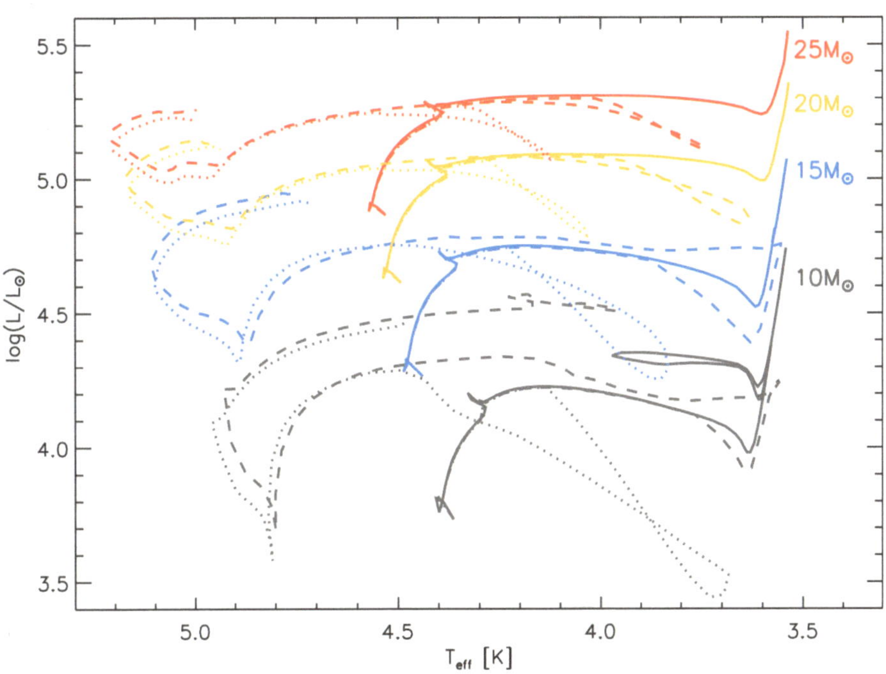

Figure 5.1. Comparison of evolutionary tracks simulating single and binary evolution from the BPASS suite of evolutionary models (e.g., Eldridge & Stanway 2016) at solar ($z = 0.14$) metallicity. The solid lines represent single evolutions for stars with initial masses of $10M_\odot$ (blue), $15_\odot$ (green), $20_\odot$ (orange), and $25_\odot$ (red). The dashed and dotted lines represent the evolution of primary stars with the same initial mass in a binary system with a main-sequence secondary (adopting a secondary-to-primary mass ratio of 0.2) and periods of 100 (dashed) and 10 (dotted) days. Note that the onset of mass transfer and the subsequent blueward evolution of the star happens much earlier for systems with shorter periods and smaller orbital separations, and that this strongly affects the lifetimes and physical properties of the RSG population.

As a result, binary evolution decreases the RSG/Wolf–Rayet ratio in a massive stellar population (e.g., Eldridge et al. 2008; Eldridge & Stanway 2009; for more discussion see Chapter 6). Another consequence of this is the presumed terminal product of the stars' evolution. Binary interaction drives down the maximum mass of a star that will undergo core collapse as an RSG, with lower-mass stars now dying as yellow supergiants or Wolf–Rayet stars (Eldridge et al. 2008, 2013). This means of stripping mass from a supernova progenitor is a popular model for explaining the blue supergiant progenitor of the Type II-pec SN 1987A (e.g., Podsiadlowski et al. 1992) and blue excesses in pre-explosion imaging of the prototypical Type IIb SN 1993J (Maund et al. 2004; for further discussion see Chapter 8). While it is difficult to disentangle the effects of binary interaction from other mechanisms that increase mass loss, such as rotation and metallicity, it is clear that it will significantly impact the evolution and death of any massive star with a close companion.

5.2 Observing RSGs in Binaries: The VV Cep and ζ Aur Systems

It is difficult to find a single comprehensive list of observed RSG binary systems. This is partly due to the rarity of RSGs, as well as the observational difficulties of detecting their binary companions. A number of RSGs or RSG candidates have proposed binary companions (including Betelgeuse and the OH/IR star VX Sgr), but the sample of RSGs with *confirmed* interacting companions is small (surprisingly so considering the key role that the RSG phase is assumed to play in the overall interacting binary fraction of massive stars) and most easily split into two subtypes, as described below.

The most famous example of an RSG with a binary companion is the VV Cephei system. The binary consists of an M-type supergiant primary, estimated between $\sim$15–20$M_\odot$, and a main-sequence B-type secondary of (roughly) comparable mass with a period of 20.344 years (e.g., Buss & Snow 1988; Bauer et al. 2007; see Figure 5.2; left panel). The large orbit size (estimated at 24.8 AU by Bennett et al. 2004) suggests that Roche lobe mass transfer has not yet occurred in its history, an assumption further supported by the eccentric orbit (e = 0.346, Wright 1977), as binaries that do undergo mass transfer tend to quickly circularize. However, the B-type companion is permanently enshrouded in circumstellar gas accreted from the *wind* of the M supergiant, completely veiling its UV spectrum.

Spectra of VV Cep show hydrogen Balmer line and continuum emission that varies in flux (sometimes strong enough to expose extremely high-order lines and dominate the spectrum blueward of $\sim$3700Å; Bennett & Bauer 2015), and the radial velocities of the Balmer lines are consistent with formation near the hot companion rather than the RSG itself. The UV spectrum also displays Fe II emission lines with significant velocity broadening ($\sim$300 km s^{-1}) that are likely pumped by Lyman series hydrogen emission. The high velocities support the model that the line-forming region is localized in the gas being accreted onto the companion, but it is unclear whether the emission is powered by gravitational energy or by the recombination of circumstellar hydrogen that is being photoionized by the B star (Bennett & Bauer 2015; Pollmann et al. 2017). VV Cep is an eclipsing binary system,

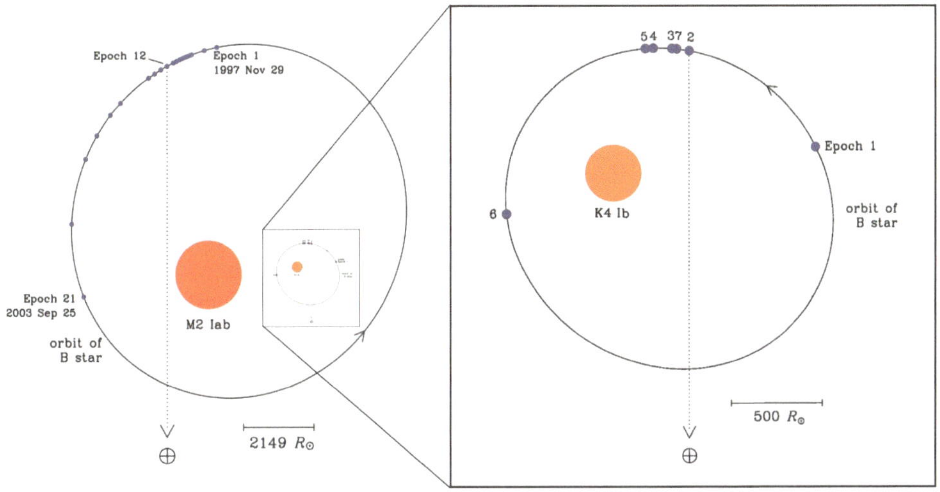

Figure 5.2. Reproduced from Bauer et al. (2007; left) and Bennett et al. (1996; right); orbits of the prototype RSG binary systems VV Cep (left) and ζ Aur (right) drawn to scale, with ζ Aur enlarged on the right for clarity. The positions of the main-sequence B-type secondaries are illustrated at various observation epochs, and the line of sight to Earth for each system is indicated by a dashed line.

with the hot companion obscured for approximately a year and a half of the period (Bauer et al. 2007).

Observations of other similar systems have given rise to a loosely defined class of "VV Cep type" systems that share several key observational characteristics with VV Cep itself: the presence of Balmer emission (often defined simply as the detection of Hα) and visual [Fe II] emission in a composite spectrum generally consisting of an M supergiant and a B-type companion (Bidelman 1954; Cowley 1969; Buss & Snow 1988). They generally have periods of 10–100 years, although α Sco is a notable exception, with an inferred period of >800 years. Most of the systems in this class are *not* eclipsing binaries, although this is not universally true (see, for example, WY Gem and AZ Cas; Ashbrook 1956; Cowley 1969; Buss & Snow 1988). The permanently enshrouded nature of the secondary B star in VV Cep is also not typical of the class, although there is evidence of similar gas accretion in these systems and UV observations of the (non-eclipsing) VV Cep binary KQ Pup show a similar spectrum to VV Cep itself, with a host of Fe II lines attributed to formation in the supergiant wind through ionization by the B star (e.g., Cowley et al. 1977; Altamore et al. 1982; Muratorio et al. 1992).

Like VV Cep, ζ Aur is the eponymous prototype star of its binary class and is comprised of a cool supergiant primary with a B-type secondary in a highly eccentric orbit ($e \sim 0.4$) with a period of ~2.66 years (Bennett et al. 1996; see Figure 5.2, right). Unlike the VV Cep class, the ζ Aur class—which also includes the binaries 31 Cyg and 32 Cyg—has K- rather than M-type supergiant primaries, orbital periods an order of magnitude shorter (~1-10 years), and a lack of the characteristic Balmer and [Fe II] emission (e.g., Eaton et al. 2008). The class is also *by definition* eclipsing, with non-enshrouded B stars that fully emerge before and after the eclipse phase (e.g., Wright 1970; Cowley et al. 1969; Bauer & Bennett 2014). Interestingly, these stars are sometimes designated as "Algol-type detached binaries," although they are not Algol variables in the traditional sense: the cool stars in these systems are the more massive of the pair and the orbital eccentricities suggest that there has been no previous instance of Roche lobe overflow, unlike the standard semi-detached Algol systems.

Phenomenologically, the VV Cep and ζ Aur binary systems are clearly quite similar; the primary distinction is one of scale, which in turn produces more interaction between the B secondary and wind material in the case of the VV Cep binaries with their more extended M supergiant winds. The lack of evidence for any true mass transfer (at least in the classical VV Cep and ζ Aur binary systems) suggests that the supergiant primaries can be effectively modeled as single stars from an evolutionary standpoint (Eaton & Shaw 2007). However, the ζ Aur binaries and eclipsing members of the VV Cep class do offer a rare observational advantage for studying the atmospheres of RSGs at very close radii by observing the B star companions as they enter and then emerge from eclipse (e.g., Bauer et al. 2007; Bauer & Bennett 2014).

22 Vul is also identified as a ζ Aur-type binary, but is set apart from the rest of the class in several key ways: its primary is classified as a G4 I yellow supergiant rather than an RSG, and it has a shorter period (249 days) and a circularized orbit. These properties suggest that the system likely experienced Roche lobe overflow and subsequent case B mass transfer in the past (Eaton & Shaw 2007). It is unclear what evolutionary relationship 22 Vul has to the ζ Aur or VV Cep binaries, and whether this previous mass transfer happened earlier in the yellow supergiant phase or even during a preceding RSG phase. It is interesting to note that Eaton & Shaw (2007) determine a luminosity of $\log(L/L_\odot)=3.3$ for the G4 I star; while this luminosity is low for a single yellow supergiant, it is surprisingly consistent with the predicted evolution of a $\sim$8-10$M_\odot$ binary primary that begins to undergo mass transfer shortly after leaving the main sequence (see, for example, the predictions of the short-period BPASS models in Figure 5.1). This could potentially mark 22 Vul as a rare and valuable observational example of a massive star that has undergone case B mass transfer.

Finally, it is worth mentioning the unusual case of V838 Mon. This star was first discovered when it underwent an outburst event in January-April 2002 and is now rapidly evolving (e.g., Bond et al. 2003; Tylenda 2005; Tylenda et al. 2011). There is still no consensus regarding the nature of the progenitor star or a physical explanation for the outburst event. However, observations following the 2002 eruption identified V838 Mon as a binary system with a very cool (late M- or even L-type) primary and a B3 V companion (Munari et al. 2002, 2007). Some results support the classification of the primary as a cool supergiant (Munari et al. 2005; Tylenda et al. 2011). Beginning in 2004, narrow [Fe II] emission lines were identified in the V838 Mon spectrum (Barsukova et al. 2006). In autumn of 2006 the system experienced an "eclipse-like" event, with the B3 V companion temporarily disappearing from the spectrum, the [Fe II] lines growing stronger, and Balmer series lines appearing in emission (Munari et al. 2007). During a second eclipse-like event in 2007, the [Fe II] emission weakened and the B3 V slowly declined, eventually disappearing from the spectrum by 2009 (Tylenda et al. 2011). Combined, the observed properties of V838 Mon between 2002-2006 in particular bear a striking similarity to those associated with VV Cep-type binaries; however, it is still unclear whether the binary nature of the system is related to the physical mechanism that drove its 2002 outburst. Interestingly, V838 Mon also displays circumstellar SiO maser emission (Deguchi et al. 2005; see Section 4.3).

5.3 Extreme Binary Scenarios for RSGs

5.3.1 X-Ray Binaries

Logically, one would expect to detect RSGs as donor stars in high-mass X-ray binaries (HMXBs); as described above, their large radii should increase the likelihood of case B or case C (common envelope) evolution while in binaries. In an extreme common envelope or merger scenario, an RSG + neutron star HMXB is also the predicted progenitor of a Thorne–Żytkow object (see below). RSGs and other evolved massive donor stars in HMXBs also represent a compelling progenitor model for dual compact object binaries, which are interesting in their own right. The merger of two neutron stars in a binary (which, presumably, were likely RSGs at previous points in their evolution due to the predicted initial mass range of NS progenitors) was recently confirmed as a short-duration gamma-ray burst progenitor scenario and as a source of gravitational waves (e.g., Abbott et al. 2017a, 2017b).

While there are a handful of massive X-ray binaries with post-main-sequence donor stars (including the Wolf–Rayet + black hole binaries IC 10 X-1, NGC 300 X-1, and M101 X-1, e.g., Bauer & Brandt 2004; Silverman & Filippenko 2008; Crowther et al. 2010; Liu et al. 2013; Binder et al. 2011, 2015), the evidence for RSG donor stars in X-ray binaries is much more limited. Observations of several ultraluminous X-ray sources (ULXs)—point-like sources with X-ray luminosities $\gtrsim 10^{39}$ ergs s^{-1} (Feng & Soria 2011)—have suggested that they may consist of RSG donor stars and massive stellar or intermediate mass black holes. These include near-infrared spectra of three presumed counterparts to ULXs (Heida et al. 2015, 2016) and photometry consistent with RSGs in seven more (López et al. 2017). However, it is difficult to confirm that these optical/IR counterparts are indeed associated with the ULXs. It is worth noting that an RSG in an X-ray binary is expected to have an observed color and spectrum that is relatively consistent with single RSGs (as irradiation from the compact companion is expected to be minimal), although the system as a whole could appear bluer than a single RSG in most color indices due to optical contributions from an accretion disk (e.g., Copperwheat et al. 2005; Heida et al. 2014). Phillips & Podsiadlowski (2002) also demonstrated that irradation *pressure* from a compact companion in an X-ray binary can deform the shape of the donor star's Roche lobe (without strongly irradiating the donor star itself), thus impacting the mass-transfer geometry and potentially decreasing the likelihood of significant accretion in an RSG + compact object binary scenario.

5.3.2 Binary Supernova Progenitors

RSGs in binary or merger scenarios have been proposed as the progenitor model for several well-known supernovae. The most famous of these is SN 1987A. Classified as a subluminous and peculiar Type II supernova in the Large Magellanic Cloud (Feast et al. 1987; Kirshner et al. 1987; Herald et al. 1987), most supernovae of this type are associated with RSGs; however, the progenitor was identified in pre-explosion observations as the B3 I blue supergiant Sk −69°202 (Sonneborn et al. 1987). The SN 1987A remnant appears as a distinctive "triple-ring" nebula, made up of material that was aspherically ejected from the progenitor ~20,000 years prior to its explosion (Wampler et al. 1990; Crotts & Heathcote 1991; Burrows et al. 1995; see Figure 5.3, left panel).

Progenitor models seeking to reproduce the supernova and ejecta morphology have invoked a binary scenario in which an RSG ejects mass in a merger with a companion. This occurs when the RSG is at or near its maximum radius, stripping the outer layers and causing a leftward evolution on the HRD as a result (e.g., Hillebrandt & Meyer 1989; Podsiadlowski et al. 1992). This scenario is also able to reproduce the observed asymmetries and composition of the remnant (e.g., Chevalier & Soker 1989; Morris & Podsiadlowski 2007; Menon & Heger 2017; see Figure 5.3, right). Similar nebulae have been observed around other blue supergiants and luminous blue variables (e.g., Brandner et al. 1997; Smartt et al. 2002; Smith et al. 2013; Gvaramadze et al. 2015), suggesting that this may be a typical mass-stripping mechanism in RSG binaries.

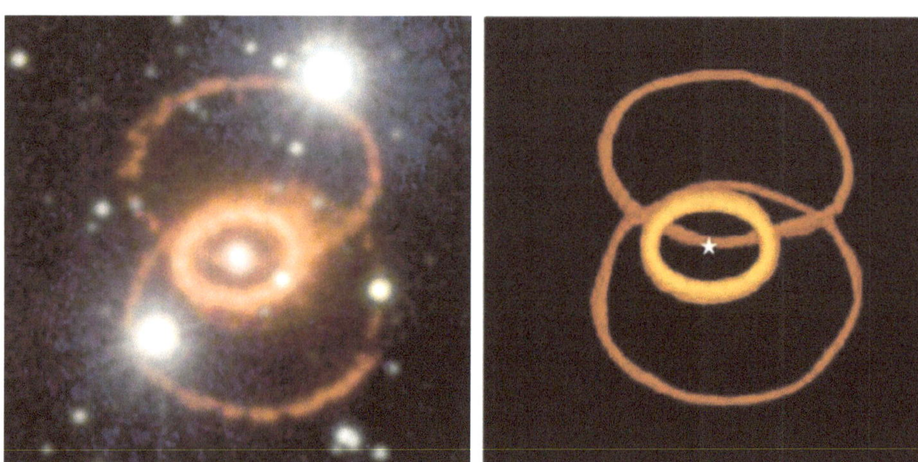

Figure 5.3. Left: courtesy of ESA/Hubble under a Creative Commons Attribution 4.0 International license. Hubble Space Telescope image of the SN 1987A remnant in February of 1994 with the Wide Field and Planetary Camera 2, showing the three rings of mass ejected by the progenitor system ~20,000 years prior to explosion. The narrow Hα + [N II] (656 nm) filter is shown in red, while the Johnson I (814 nm) and U (336nm) filters are shown in green and blue, respectively. Right: from Morris & Podsiadlowski (2007; reprinted with permission from AAAS); simulated Hα (656nm) emission for their three-dimensional smooth-particle hydrodynamics simulations of a 20$M_\odot$, 1500$R_\odot$ RSG progenitor evolving through a common envelope phase, binary merger, mass ejection, and subsequent wind-driven evolution.

Another supernova with a presumed RSG binary progenitor is SN 1993J, which occurred in the nearby galaxy M81. The supernova spectrum was initially that of a Type II supernova (displaying hydrogen in the spectrum), but later evolved to resemble a Type Ib (lacking hydrogen); this became the prototypical example of the Type IIb class of supernova. The explosive properties of the supernova were consistent with a $\sim$4-5$M_{\odot}$ He core and a low-mass extended envelope. Models seeking to explain this invoked a binary scenario (e.g., Nomoto et al. 1993; Podsiadlowski et al. 1993; Woosley et al. 1994), with a $\sim$15$M_{\odot}$ primary becoming an RSG, filling its Roche lobe, and losing $\sim$10$M_{\odot}$ during classic case B mass transfer onto a lower-mass companion. Studies of the copious pre-explosion imaging available for M81 showed that the spectral energy distribution of the progenitor was best modeled by combining that of a G-type to K-type supergiant with a blue component (Aldering et al. 1994). Two decades of rigorous photometric and spectroscopic follow-up campaigns focused on searching for signs of the surviving secondary star. This work ultimately identified the signature of a B-type companion to the SN 1993J progenitor, suggesting a model in which a K-type RSG progenitor with a smaller blue secondary underwent strong mass transfer $\sim$8000 years before the explosion (e.g., Van Dyk et al. 2002; Maund et al. 2004; Weiler et al. 2007; Smartt 2009; Fox et al. 2014).

For more discussion of RSGs and their role as supernova progenitors, see Chapter 8.

5.3.3 Thorne–Żytkow Objects

Models of common envelope evolution in massive stars frequently invoke the RSG phase as a key component. In these scenarios, the RSG fills its Roche lobe and begins mass transfer onto a close companion, which then fills its own Roche lobe and ultimately results in the close orbit of the primary and secondary cores within a shared stellar envelope (e.g., Taam & Sandquist 2000; Podsiadlowski 2001; Ivanova et al. 2001). One expected outcome of this evolution is the eventual merger of the two stars; however, the end product is strongly dependent on the evolutionary phase of the binary components when they begin interacting, with scenarios ranging from mergers during the main sequence to compact object mergers (e.g., Kochanek et al. 2014; de Mink et al. 2014; de Mink & Belczynski 2015).

Perhaps the most extreme end product of common envelope evolution in RSGs is the Thorne–Żytkow object (TŻO). Originally predicted in the mid-1970's, TŻOs consist of a neutron star core surrounded by a large, cool, diffuse envelope (Thorne & Żytkow 1975, 1977). Massive TŻOs are supported primarily by thermonuclear reactions at the base of the convective envelope, with a small contribution ($\sim$5%) from gravitational accretion onto the neutron star (Thorne & Żytkow 1977; Biehle 1991). A TŻO is expected to form as the end result of a massive binary composed of an RSG and a neutron star. Taam et al. (1978) proposed that a TŻO could form following an HMXB phase, with the expanding RSG envelope eventually leading to common envelope evolution, while Leonard et al. (1994) propose that this in-spiral and merger could occur when a newly formed neutron star receives a supernova "kick" velocity in the direction of its RSG companion and becomes embedded. Either scenario can lead to the in-spiral of the neutron star, which tidally disrupts the existing core and eventually settles at the center of the supergiant envelope (Tout et al. 2014).

Observational evidence of the existence of TŻOs remained elusive for nearly 40 years—TŻOs are outwardly almost identical to RSGs and can only be distinguished from their more common doppelgängers by their atypical chemical abundances. These abundances arise as a result of the interrupted-rapid-proton process (irp-process), made possible by the high temperatures at the neutron star surface combined with the completely convective cool supergiant envelope. This produces excesses of elements including Rb, Sr, Y, Ze, and Mo that can be observed via atypically strong absorption lines in the RSG spectrum (Wallace & Woosley 1981; Cannon 1993; Biehle 1994). TŻO spectra should also display an excess of ^{7}Li, a result of the ^{7}Be-transport mechanism's dependence on the star's internal structure (e.g., Cameron 1955; Podsiadlowski et al. 1995). Feedback from TŻOs could thus also be important when considering the observed abundances of lithium and some more massive proton-rich nuclei (e.g., Cannon 1993).

In 2014, the first strong candidate for observational evidence of a TŻO was detected in the SMC; the presumed SMC RSG HV 2112 was found to have atypical line ratios that set it apart from the rest of the RSG population but agreed well with predictions for TŻO spectra (Levesque et al. 2014; see Figure 5.4, top). HV 2112 also showed an excess amount of Ca I, consistent with predictions of the TŻO formation process (Tout et al. 2014), and has a proper motion consistent with membership in the SMC (Worley et al. 2016); combined, these suggest that it is a true SMC supergiant rather than a polluted foreground giant or dwarf in the Milky Way. HV 2112 is also strongly photometrically and spectroscopically variable (e.g., Payne-Gaposchkin & Gaposchkin 1966; Reid & Mould 1990; Levesque et al. 2014), and displays a variable spectrum of strong Balmer emission with an unusual Balmer "increment" (Hδ, rather than Hα, is the strongest feature; see Figure 5.4, bottom panel); however, it does not show any of the signs of a hot companion seen in the VV Cep or ζ Aur variables (see above). The variable emission spectrum of HV 2112 is most comparable to the spectral variations see in lower-mass Mira variables and is attributed to a similar formation mechanism, produced by non-LTE radiative transfer effects in the hydrogen lines from formation in an atmosphere that is propagating shocks generated by pulsational instabilities (Castalez et al. 2000; Levesque et al. 2014).

Many unanswered questions still remain regarding TŻOs, mostly focused on their stability, lifetimes, and predicted final products. van Paradijs (1995) predicts that 1 in 100 Galactic RSGs could instead be a TŻO; however, a careful prediction of the TŻO rate must take into account several key factors. The metallicity dependence of TŻO formation could be significant, with the strengths of massive stellar winds depending on abundance and impacting the orbital radii and merger rates of binaries (e.g., Linden et al. 2010; Douna et al. 2015); from this we would expect TŻO formation to be more common at low metallicity. The stability and end product of the TŻO model are also unclear. If there is neutrino-driven runaway accretion onto

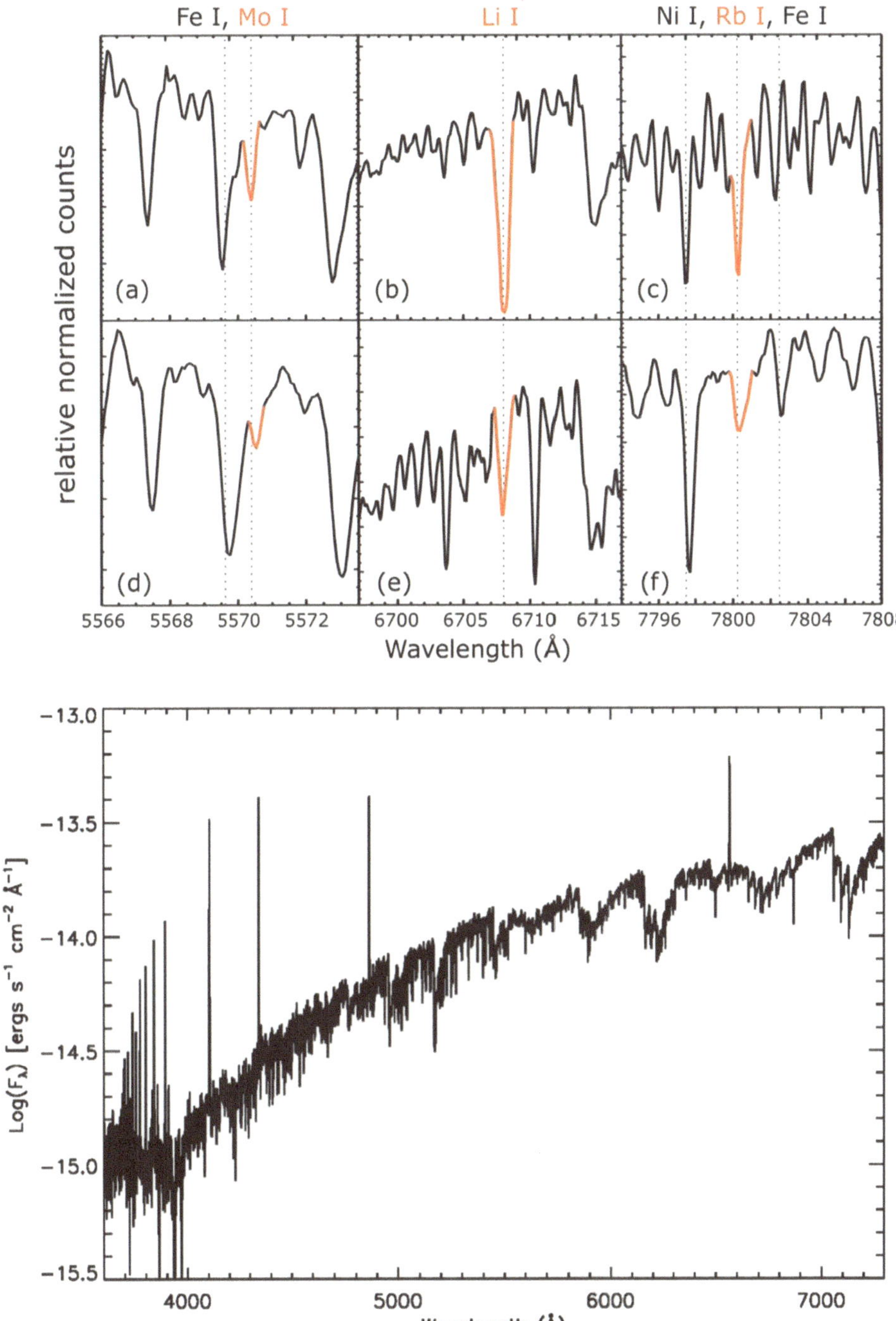

Figure 5.4. Reproduced from Levesque et al. (2014). Top: echelle spectra of HV 2112 (top) and SMC 005092 (bottom), a typical RSG. The plotted features include Fe I λ5569.62 and Mo I λ 5570.40 (a,d); Li I λ6707.97 (b, e); and Ni I λ7797.58, Rb I λ7800.23, and Fe I λ7802.47 (c,f); wavelengths of the absorption features of interest are marked as vertical dashed lines and features predicted to be enhanced in TŻOs are marked in red. Bottom: optical spectrum of HV 2112, showing the strong Balmer emission features.

the neutron star during or following the in-spiral, it will quickly exceed its maximum mass and become a black hole, rendering the TŻO model unfeasible (e.g., Fryer et al. 1999). Alternately, if the accretion rate is low and the death of a TŻO is ultimately caused by the core's transition from a neutron star to a black hole, TŻO lifetimes could dramatically exceed those of RSGs, extending to billions of years. A more likely terminal model for TŻOs is the simple shedding of the envelope through strong cool supergiant mass loss, which would correspond to a significantly shorter lifetime and leave behind a lone neutron star that has been significantly spun down. This scenario has been proposed as the origin of the slowly rotating X-ray source 1E161348-5055 in the young supernova remnant RCW 103 (Liu et al. 2015).

The definitive detection of a TŻO would provide the first direct evidence for a completely new stellar interior model and unique nucleosynthesis processes that offer a channel for lithium and heavy element production in our universe. TŻOs would also serve as proof of a theoretically predicted end stage of massive binary evolution. Combined with a clearer understanding of the processes that constrain their lifetimes, the TŻO population could serve as an important test case for understanding the nature and frequency of common envelope evolution and binary mergers involving RSGs.

References

Abbott, B. P., Abbott, R., Abbott, T. D., et al. 2017a, ApJL, **848, 12**

Abbott, B. P., Abbott, R., Abbott, T. D., et al. 2017b, ApJL, **848, 13**

Aldering, G., Humphreys, R. M.., & Richmond, M. 1994, AJ, **107, 662**

Altamore, A., Giangrande, A., & Viotti, R. 1982, A&AS, **49, 511**

Ashbrook, J. 1956, Harvard Obs. Announcement Card No. 1340

Barsukova, E., Goranskij, V., Abolmasov, P., & Fabrika, S. 2006, ATel, **803, 1**

Bauer, F. E., & Brandt, W. N. 2004, ApJ, **601, L67**

Bauer, W. H., Bennett, P. D., & Brown, A. 2007, ApJS, **171, 249**

Bauer, W. H., & Bennett, P. D. 2014, ApJS, **211, 27**

Bennett, P. D., & Bauer, W. H. 2015, in Astrophysics and Space Science Library, Giants of Eclipse: The ζ Aurigae Stars and Other Binary Systems (Vol. 408; Basel: Springer), 85

Bennett, P. D., Brown, A., Fawcett, S. M., Yang, S., & Bauer, W. H. 2004, in ASP Conf. Ser. 318, Spectroscopically and Spatially Resolving the Components of Close Binary Stars, ed. R. W. Hilditch, H. Hensberge, & K. Pavlovski (San Francisco, CA: ASP), 222

Bennett, P. D., Harper, G. M., Brown, A., & Hummel, C. A. 1996, ApJ, **471, 454**

Bidelman, W. P. 1954, ApJS, **1, 175**

Biehle, G. W. 1991, ApJ, **380, 167**

Biehle, G. T. 1994, ApJ, **420, 364**

Binder, B., Williams, B. F., Kong, A. K. H., et al. 2011b, ApJL, **739, L51**

Binder, B., Gross, J., Williams, B. F., & Simons, D. 2015, MNRAS, **451, 4471**

Bond, H. E., Henden, A., Levay, Z. G., et al. 2003, Natur, **422, 405**

Brandner, W., Chu, Y., Eisenhauer, F., Grebel, E. K., & Points, S. D. 1997, ApJL, **489, L153**

Burrows, C. J., Krist, J., Hester, J. J., et al. 1995, ApJ, **452, 680**

Buss, R. H., & Snow, T. P. 1988, ApJ, **335, 331**

Cameron, A. G. W. 1955, ApJ, **121, 144**

Cannon, R. C. 1993, MNRAS, **263, 817**
Castalez, M. W., Luttermoser, D. G., Caton, D. B., & Piontek, R. A. 2000, AJ, **120, 2627**
Chevalier, R. A., & Soker, N. 1989, ApJ, **341, 867**
Copperwheat, C., Cropper, M., Soria, R., & Wu, K. 2005, MNRAS, **362, 79**
Cowley, A. P. 1969, PASP, **81, 297**
Cowley, A. P., Hutchings, J. B., & Popper, D. M. 1977, PASP, **89, 882**
Crotts, A. P., & Heathcote, S. R. 1991, Natur, **350, 683**
Crowther, P. A., Barnard, R., Carpano, S., et al. 2010, MNRAS, **403, L41**
Deguchi, S., Matsunaga, N., & Fukushi, H. 2005, PASJ, **57, L25**
de Mink, S. E., Sana, H., Langer, N., Izzard, R. G., & Schneider, F. R. N. 2014, ApJ, **782, 7**
de Mink, S. E., & Belczynski, K. 2015, ApJ, **814, 58**
Douna, V. M., Pellizza, L. J., Mirabel, I. F., & Pedrosa, S. E. 2015, A&A, **579, 44**
Eaton, J. A., Henry, G. W., & Odell, A. P. 2008, ApJ, **679, 1490**
Eaton, J. A., & Shaw, F. G. 2007, AJ, **133, 2669**
Eldridge, J. J., Fraser, M., Smartt, S. J., Maund, J. R., & Crockett, R. M. 2013, MNRAS, **436, 774**
Eldridge, J. J., Izzard, R. G., & Tout, C. A. 2008, MNRAS, **384, 1109**
Eldridge, J. J., & Stanway, E. R. 2009, MNRAS, **400, 1019**
Eldridge, J. J., & Stanway, E. R. 2016, MNRAS, **462, 3302**
Feast, M. W., Winkler, H., Catchpole, R. et al. 1987, IAU Circ, **4320, 1**
Feng, H., & Soria, R. 2011, NewAR, **55, 166**
Fox, O. D., Azalee Bostroem, K., Van Dyk, S. D., et al. 2014, ApJ, **790, 17**
Fryer, C. L., Woosley, S. E., & Hartmann, D. H. 1999, ApJ, **526, 152**
Garmany, C. D., Conti, P. S., & Massey, P. 1980, ApJ, **242, 1063**
Gvaramadze, V. V., Kniazev, A. Y., Bestenlehner, J. M., et al. 2015, MNRAS, **454, 219**
Heida, M., Jonker, P. G., Torres, M. A. P., et al. 2014, MNRAS, **442, 1054**
Heida, M., Jonker, P. G., Torres, M. A. P., et al. 2016, MNRAS, **459, 771**
Heida, M., Torres, M. A. P., Jonker, P. G., et al. 2015, MNRAS, **453, 3510**
Herald, D., McNaught, R. H., Morel, M., et al. 1987, IAU Circ, **4317, 1**
Hillebrandt, W., & Meyer, F. 1989, A&A, **219, L3**
Ivanova, N., Podsiadlowski, P., & Spruit, H. 2001, in ASP Conf. Ser. 229, Evolution of Binary and Multiple Star Systems, ed. Ph. Podsiadlowski et al. (San Francisco, CA: ASP), 261
Kirshner, R. P., Sonneborn, G., Cassatella, A., et al. 1987, IAU Circ, **4435, 1**
Kochanek, C. S., Adams, S. M., & Belczynski, K. 2014, MNRAS, **443, 1319**
Lamers, H. J. G. L. M. & Levesque, E. M. 2017, Understanding Stellar Evolution (Bristol: IOP)
Leonard, P. J. T., Hills, J. H., & Dewey, R. J. 1994, ApJL, **423, L19**
Levesque, E. M., Massey, P., Zytkow, A. N., & Morrell, N. 2014, MNRAS, **443, L94**
Linden, T., Kalogera, V., Sepinsky, J. F., et al. 2010, ApJ, **725, 1984**
Liu, J.-F., Bregman, J. N., Bai, Y., Justham, S., & Crowther, P. 2013, Natur, **503, 500**
Liu, X. W., Xu, R. X., van den Heuvel, E. P. J., et al. 2015, ApJ, **799, 233**
López, K. M., Heida, M., Jonker, P. G., et al. 2017, MNRAS, **469, 671**
Maund, J. R., McNaught, R. H., Morel, M., et al. 2004, Natur, **427, 129**
Menon, A., & Heger, A. 2017, MNRAS, in press
Morris, T., & Podsiadlowski, P. 2007, Sci, **315, 1103**
Munari, U., Corradi, R. L. M., Henden, A., et al. 2007, A&A, **474, 585**
Munari, U., Desidera, S., & Henden, A. 2002, IAU Circ., **8005, 2**

Munari, U., Henden, A., Vallenari, A., et al. 2005, A&A, **434, 1107**
Muratorio, G., Viotti, R., Friedjung, M., Baratta, G. B, & Rossi, C. 1992, A&A, **258, 423**
Nomoto, K., Suzuki, T., Shigeyama, T., et al. 1993, Natur, **364, 507**
Payne-Gaposchkin, C., & Gaposchkin, S. 1966, SCoA, **9, 1**
Phillips, S. N., & Podsiadlowski, P. 2002, MNRAS, **337, 431**
Podsiadlowski, P. 2001, in ASP Conf. Ser. 229, Evolution of Binary and Multiple Star Systems, ed. Ph. Podsiadlowski et al. (San francisco, CA: ASP), 239
Podsiadlowski, P., Cannon, R. C., & Rees, M. J. 1995, MNRAS, **274, 385**
Podsiadlowski, P., Hsu, J. J. L., Joss, P. C., & Ross, R. R. 1993, Natur, **364, 509**
Podsiadlowski, P., Joss, P. C., & Hsu, J. J. L. 1992, ApJ, **391, 246**
Pollmann, E., Vollmann, W., & Bennett, P. D. 2017, IBVS, **6198, 1**
Reid, N., & Mould, J. 1990, ApJ, **360, 490**
Sana, H., de Mink, S. E., de Koter, A., et al. 2012, Sci, **337, 444**
Smartt, S. J. 2009, ARA&A, **47, 63**
Smartt, S. J., Lennon, D. J., Kudritzki, R. P., et al. 2002, A&A, **391, 979**
Smith, N., Arnett, W. D., Bally, J., Ginsburg, A., & Filippenko, A. V. 2013, MNRAS, **429, 1324**
Silverman, J. M., & Filippenko, A. V. 2008, ApJL, **678, L17**
Sonneborn, G., Altner, B., & Kirshner, R. P. 1987, ApJ, **323, 35**
Taam, R. E., Bodenheimer, P., & Ostriker, J. P. 1978, ApJ, **222, 269**
Taam, R. E., & Sandquist, E. L. 2000, ARA&A, **38, 113**
Thorne, K. S., & Zytkow, A. N. 1975, ApJL, **199, L19**
Thorne, K. S., & Zytkow, A. N. 1977, ApJ, **212, 832**
Tout, C. A., Zytkow, A. N., Church, Ross P., et al. 2014, MNRAS, **445, L36**
Tylenda, R., Kaminski, T., Schmidt, M., Kurtev, R., & Tomov, T. 2011, A&A, **532, 138**
Tylenda, R. 2005, A&A, **436, 1009**
van Dyk, S. D., Garnavich, P. M., Filippenko, A. V., et al. 2002, PASP, **114, 1322**
van Paradijs, J., Spruit, H. C., van Langevelde, H. J., & Waters, L. B. F. M. 1995, A&A, **303, L25**
Wallace, R. K., & Woosley, S. E. 1981, ApJS, **45, 389**
Wampler, E. J., Wang, L., Baade, D., et al. 1990, ApJ, **362, 13**
Weiler, K. W., Williams, C. L., Panagia, N., et al. 2007, ApJ, **671, 1959**
Woosley, S. E., Eastman, R. G., Weaver, T. A., & Pinto, P. A. 1994, ApJ, **429, 300**
Worley, C. C., Irwin, M. J., Tout, C. A., et al. 2016, MNRAS, **459, L31**
Wright, K. O. 1970, VA, **12, 147**
Wright, K. O. 1977, JRASC, **71, 152**

IOP Astronomy

Emily M. Levesque

Chapter 6

Red Supergiants in and beyond the Milky Way

6.1 Identifying Red Supergiant Populations

Observationally identifying a population of RSGs seems straightforward, following the same conversion between physical properties and observables that is used to translate the Hertzsprung–Russell diagram to the color–magnitude diagram. The range of RSG effective temperatures can be mapped to a range of color indices, while their luminosities, combined with a well-known distance, can be converted into apparent magnitudes. Most efforts to identify the RSG population of a cluster or galaxy begin in exactly this manner, adopting the colors and magnitudes that are best able to distinguish cool bright stars from the surrounding population. On a color–magnitude diagram of the Magellanic Clouds, Massey (2002) identified a "strong diagonal feature" spanning from $B - V = 1.5$ and $V = 14$ to $B - V \sim 2$ and $V = 12$ that was primarily made up of member RSGs - the color ranges correspond to those expected for a population of early K- to mid-M-type RSGs, and this same RSG feature appears in color–magnitude diagrams of other massive star populations (see Figure 6.1, left). Massey & Olsen (2003) quantified color and magnitude cutoffs for identifying an RSG sample in *UBVRI* data, with $(V - R)_0 \geqslant 0.6$ and a cutoff in V such that $M_{\rm bol} < -7.0$.

While these simple color and magnitude criteria can identify RSGs, they fail to account for a critical issue in identifying distant and luminous red sources: foreground contamination. Massey (1998) first identified this problem in data from the Humphreys & Sandage (1980) study of luminous stars in M33. The distribution of blue supergiants was, as expected, clumpy across the face of the galaxy (with the massive blue stars concentrated in areas of recent star formation), but the population of apparent RSGs was much more uniformly distributed, a result of contamination from foreground red dwarfs. With similar apparent magnitudes to background RSGs in Local Group galaxies, the contamination fraction from this population is significant (Ratnatunga & Bahcall 1985 estimated a contamination fraction of 0.1 foreground dwarfs per square arcmin down to $V = 19$ for a typical

doi:10.1088/978-0-7503-1329-2ch6

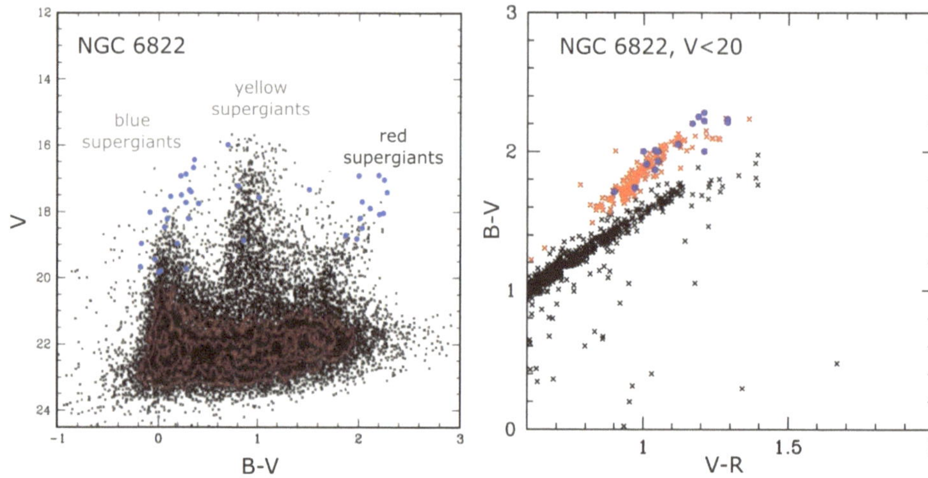

Figure 6.1. Color–magnitude and color–color diagrams of the RSG population in NGC 6822, which has a substantial foreground contaminant population due to its low Galactic latitude (b = −18.4 deg). Left: reproduced from Massey et al. (2007); color–magnitude diagram of luminous stars observed in NGC 6822, with the diagonal sequence dominated by RSGs identified. Right: reproduced from Levesque & Massey (2012); the color–color diagram of RSGs in NGC 6822, with foreground stars in black and RSGs in red, separated using the Massey (1998) criteria. In both panels, blue points represent supergiants that have been spectroscopically confirmed as members of NGC 6822.

Local Group galaxy, with a much higher fraction of 0.6 foreground dwarfs per square arcmin between $19 < V < 21$; the effect worsens at lower Galactic latitudes).

Massey (1998) presented a simple but effective means of removing foreground dwarfs from a background RSG sample. As described previously in Chapter 3, surface-gravity-dependent line-blanketing effects can be used to separate the colors of high-gravity dwarfs and low-gravity RSGs. The effects of line blanketing are increased at low surface gravities in these stars, and are particularly notable in the *B* band due to the number of weak metal lines in that regime, decreasing the *B* fluxes of stars with lower surface gravities. On a $(V - R)$ vs. $(B - V)$ color–color diagram, foreground dwarfs fall along a distinct diagonal sequence that is bluer in $(B - V)$ than their lower-gravity counterparts at the same $(V - R)$ (see Figure 6.1, right panel). The dwarf and supergiant sequences can be separated by a cutoff of $(B - V) = 1.26 \times (V - R) + 0.45$; this method is extremely effective at removing foreground dwarf contaminants from a background RSG population.

With spectroscopic observations, foreground and background samples can sometimes also be separated kinematically; however, this method requires that the motion of the foreground and background sources are both well-known. It also fails to account for the possibility of "runaway" stars, which have been identified in both Galactic and extragalactic RSG populations. Gies & Bolton (1986) estimate that a significant fraction of OB stars (10–50%) are considered runaways (with peculiar velocities of $\gtrsim$30 km s^{-1}; Blaauw 1961), likely as a consequence of dynamical evolution (see also Fujii & Portegies Zwart 2011; Oh et al. 2015), and one would logically expect to also see the evolved descendents of these runaways. A handful of

Galactic RSGs have been identified as runaways due to the presence of bow shocks, including Betelgeuse (56 km s^{-1}; Noriega-Crespo et al. 1997; Mackey et al. 2012), μ Cephei (22 km s^{-1}; Cox et al. 2012), and IRC −10414 (70 km s^{-1}, Gvaramadze et al. 2014). Extragalactically, Evans & Massey (2015) identified a runaway RSG member of M31—its 300 km s^{-1} radial velocity makes it the fastest known runaway massive star—and note other potential runaway RSG candidates in the Large Magellanic Cloud and M33 with peculiar velocities of >100 km s^{-1}. These and other potentially interesting RSGs run the risk of being wrongfully eliminated from a background population in kinematic analyses.

Foreground red giants are also potential contaminants that can stymie kinematic methods (and cannot be photometrically separated from RSGs like their higher-gravity dwarf counterparts). Giants in the Milky Way halo can have similar velocities to some Local Group galaxies, as these are due mainly to the reflex motion of the Sun. In some cases, high-quality proper motions and parallaxes can be used to address this ambiguity (see, for example, the case of TŻO candidate HV 2112; Worley et al. 2016); current efforts such as the GAIA astrometric mission will make significant strides in dynamically separating foreground and background populations (e.g., Perryman et al. 2001). However, the expected fraction of halo giant contamination is much lower than that of foreground dwarfs (see, for example, Drout et al. 2009, 2012; Neugent et al. 2012).

While RSGs and red giants stars are quite distinct from an evolutionary standpoint (see Chapter 2), they are hard to separate observationally. Even if cluster or galaxy membership can be confirmed, "super"-AGB stars can have luminosities that overlap with those of standard ~8–10$M_\odot$ RSGs (e.g., Siess 2006; Poelarends et al. 2008) and similarly lie on the Hayashi limit, making them virtually indistinguishable on Hertzsprung–Russell and color–magnitude diagrams. There is no clear photometric diagnostic for distinguishing between luminous giant and lower-mass RSG populations—the most common approach, as reflected in the Massey & Olsen (2003) criteria, is to simply adopt a conservative lower limit for $M_{\rm bol}$, but this limits the completeness of the RSG sample by filtering out the lowest-mass (and most numerous) members of the population. Optical spectral lines such as the Ca II triplet and the Hα absorption feature are sensitive to surface gravity and broader in supergiants as compared to giants (see Section 3.6), but these diagnostics fail in cool stars with emission spectra (including Mira variables, giant and supergiant interacting wind binaries—see Section 5.2—and Thorne–Żytkow objects—Section 5.3). High-resolution spectroscopic follow-up is also logistically impractical for sorting through large samples of cool stars. Similarly, the 86 GHz SiO maser line is much broader in RSGs than in Miras, an apparent consequence of the larger terminal expansion velocity that results from stronger mass loss in RSGs (Alcolea et al. 1990; Verheyen et al. 2012), but use of this effect to distinguish giants and supergiants is limited to stars with circumstellar masers.

Finally, it is also possible to occasionally misidentify even more distant background objects as RSGs based on photometry. The most common background contaminants from *BVR* photometry are luminous red galaxies, whose SEDs are dominated by their lower-mass red star populations; at larger distances globular clusters can be mistaken for RSGs in a similar fashion. These can sometimes be solved through inspection of imaging data, with galaxies appearing as extended sources, but ambiguities still remain if the background source is sufficiently distant to appear point-like or the foreground RSG has an extended circumstellar dust envelope (see, for example, Section 4.3). The red object LGGS J004527.30+413254.3 was originally classified as both an M31 RSG and a globular cluster candidate before spectroscopy confirmed its nature as a low-luminosity AGN (Dorn-Wallenstein et al. 2017).

Near- and mid-infrared criteria can also be used to identify RSG populations; these are particularly valuable when working along high-extinction sightlines. A number of massive clusters in the Milky Way with significant RSG populations are located at low Galactic latitudes and near the Galactic Center, including the Arches, Quintuplet, and Central clusters, as well as five "red supergiant clusters" (RSGC1-5; see Section 6.2). The RSGC stars were originally identified as bright K_s sources from the Two Micron All Sky Survey (2MASS) catalog, with follow-up spectra taken to confirm their stellar classification and membership. While RSGs do cluster on near-IR color magnitude diagrams such as $(J - K_s)$ vs. K_s or $(H - K_s)$ vs. K_s, they do not follow a sequence that is easily quantified photometrically. The $Q1$ parameter (where $Q1 = (J - H) - 1.8 \times (H - K_s)$; Negueruela & Schurch 2007) has been used as a tool for identifying RSGs in IR data; González-Fernández et al. (2015) combined a range of $0.1 \leq Q1 \leq 0.4$ with a minimum $K_s > 11$ and basic proper motion criteria to identify RSG candidates in the Magellanic Clouds (see Section 6.3).

Combining near- and mid-IR photometry (when available) can be a successful means of identifying extragalactic RSGs, leveraging the high mid-IR luminosities that arise as a result of significant mass loss and dust production (a trait not shared by foreground dwarfs). Britavskiy et al. (2014) defined a suite of selection criteria that can be applied to populations with a well-defined distance: M[3.6] < −9; $J - [3.6] > 1$ and $[3.6] - [4.5] < 0$; and $J - H > 0.65$ and $H - K < 0.6$ (see also Bonanos et al. 2009, 2010). These have been used to successfully identify some RSG members of Local Group galaxies including IC 1613, Pegasus, Phoenix, Sextans A, and WLM ; however, contamination by foreground giants remains a challenge (see also Britavskiy et al. 2015). Messineo et al. (2012) combine the $Q1$ parameter with their newly established $Q2 = (J - K_s) - 2.69 \times (K_s - [8.0])$ parameter to identify massive star populations; a parameter space of $0.1 < Q1 < 0.5$ and $0.5 < Q2 < 1.5$ encompasses 42% of known Galactic RSGs (see also Messineo et al. 2014a, 2014b).

6.2 RSGs in the Milky Way

The best-studied RSGs in the Milky Way are those with probable memberships in OB associations or clusters with known distances that lie far from the Galactic plane. These were first cataloged beginning in the 1970s (e.g., Humphreys 1970; Humphreys & McElroy 1984; Garmany & Stencel 1990), and are relatively rare—the Per OB1 association contains ~20 RSGs, likely produced from several star formation episodes spanning ~10 Myr (Slesnick et al. 2002), while the coeval cluster NGC 7419 contains only five RSGs with an age of ~14 ± 2 Myr (Caron et al.

Figure 6.2. Courtesy of NASA. False-color composite image of RSGC1 (Figer et al. 2006) from 2MASS; J band is shown in blue, H in green, and K_s in red.

2003; Marco & Negueruela 2013). Their known distances and minimal Galactic extinction make these RSGs prime targets for detailed spectroscopic observations (see Levesque et al. 2005; Gazak et al. 2014).

In the past decade, a significant RSG population has also been discovered toward the Galactic Center. Five RSG clusters—known as RSGC1-5—are localized at the base of the Scutum-Crux arm, near its interface with the Galactic Bar. These clusters contain at least 76 spectroscopically confirmed RSGs, making this the most RSG-rich region currently known in the Milky Way (e.g., Figer et al. 2006; Davies et al. 2007; Clark et al. 2009; Negueruela et al. 2010, 2011; Figure 6.2) and indicating a massive recent burst of star formation in the region. More recently, Messineo et al. (2016) identified 58 new likely RSGs in the inner Galaxy. Unlike their counterparts at higher latitudes, these RSGs have, out of necessity, been primarily studied in the IR and radio out of necessity (see also Verheyen et al. 2012). The population of low-latitude RSGs is expected to grow further with new IR-focused observing campaigns (see Chapter 9).

Even combining stars in well-studied OB associations and open clusters with the recent glut discovered toward the Galactic Center, only a few hundred confirmed RSGs have been identified in the Milky Way. This is at odds with population predictions, which suggest there should be several thousand Milky Way RSGs (e.g., Garwood & Jones 1987; Gehrz 1989; Messineo et al. 2012).

6.3 Extragalactic RSG Populations

RSGs are luminous enough to be studied well beyond our own Galaxy, with photometry and spectroscopy of individual RSGs possible throughout and beyond the Local Group. These extragalactic massive star populations are immensely valuable for studying stellar populations in other galaxies, particularly with respect to how metallicity and other ISM properties impact stellar evolution.

The Large and Small Magellanic Clouds (LMC and SMC) are our nearest galactic neighbors and serve as rich resources for studying extragalactic stellar populations. Massey (2002) and Massey & Olsen (2003) originally identified a substantial population of RSGs in the Large and Small Magellanic Clouds. These were later followed up with optical spectrophotometry beginning with a representative member sample of 36 LMC RSGs and 39 SMC RSGs (Levesque et al. 2006). Neugent et al. (2012) further winnowed down the likely RSG member population in the LMC to 505 stars using proper motions and the galaxy's large systemic radial velocity. This work has most recently been extended to a large-scale spectrophotometric survey covering hundreds of LMC and SMC RSGs. Using near-IR criteria, González-Fernñandez et al. (2015) also identified a sample of cool RSG candidates; optical spectroscopic follow-up yielded a final sample of 123 LMC RSGs and 160 SMC RSGs, though it is worth noting that they expand their definition of RSGs to cool supergiants with spectral types of G0 or later (for further discussion see Section 6.4).

Moving beyond the Magellanic Clouds, the photometric (*UBVRI*) Local Group Galaxy Survey (Massey et al. 2006, 2007) was a key starting point for identifying RSGs (and other massive star populations) in the remaining galaxies of the Local Group. In the spiral galaxies M31 and M33, follow-up spectroscopic work identified true RSG members through a combination of the color–color method and radial velocities, and a combination of spectra and photometric diagnostics made it possible to determine the effective temperatures and bolometric luminosities of the RSG populations and place them on the H-R diagram (Massey et al. 2009; Drout et al. 2012; Massey & Evans 2016). More recently, this work has been extended to spectrophotometric follow-up of Local Group Galaxy Survey RSGs identified in nearby dwarf galaxies; Levesque & Massey (2012) obtained spectra of 16 RSGs in NGC 6822 and 11 RSGs in WLM. Britavskiy et al. (2014, 2015) used optical spectroscopy to follow up RSG candidates identified in Local Group dwarf irregular galaxies using their near- and mid-IR photometry criteria. Their sample included spectra of seven RSGs in Sextans A, four in WLM (also previously observed in Levesque & Massey 2012), two in IC 1613, two in Pegasus, and one in Phoenix.

In recent years RSGs have also been studied with near-IR spectra, pushing observations of individual massive stars well beyond the Local Group. Gazak et al. (2015) observed *J*-band spectra of 27 RSGs in NGC 300, at a distance of 1.88 Mpc, while Patrick et al. (2017) obtained *J*-band spectra of 18 RSGs in NGC 55, at a distance of 2.35 Mpc. Closer to home, *J*-band spectra have been obtained for RSGs within the Milky Way (Gazak et al. 2014), the Magellanic Clouds (Davies et al. 2013, 2015; Patrick et al. 2016), and NGC 6822 (Patrick et al. 2015). While these targets were mainly selected using optical photometry, the near-IR spectroscopic follow-up offers an intriguing alternative for how extragalactic RSGs can be observed and utilized. A satisfactory effective temperature diagnostic for RSGs has not yet been developed from near-IR spectra (with current techniques over-estimating temperatures and unable to reproduce predicted metallicity effects; see Section 3.2 and Section 6.4); however, these studies have demonstrated the substantial potential of using near-IR spectra of RSGs to determine their host galaxies' metallicities, with inferred metallicities from RSG spectra agreeing extremely well with those determined from blue supergiant spectra. In the coming decade, near-IR spectra of extragalactic RSGs (and, at greater distances, RSG-dominated clusters; see Lardo et al. 2015) could serve as powerful probes of extragalactic stellar metallicities (for further discussion see Section 9.2).

6.4 Metallicity Effects in RSGs

RSG populations are strongly sensitive to the metallicities of their host environments. With their young ages, the metallicity of a massive star population can be taken as representative of its natal environment metallicity. This same metallicity strongly affects the evolution of massive stars throughout their lifetimes; these effects have been both theoretically predicted and observationally confirmed in studies of extragalactic RSG populations. As demonstrated above, spectroscopic observations of individual RSGs now span a broad range of host environments, from extremely metal-poor galaxies such as IC 1613 ($\log(O/H)+12{\sim}7.62$; Lee et al. 2003) and WLM ($\log(O/H)+12{\sim}7.83$; Lee et al. 2005, Bresolin et al. 2006) to the super-solar metallicities measured in regions of M31 ($\log(O/H)+12 \sim 9.3$; Sanders et al. 2012).

An immediate effect of metallicity can be seen in its impact on the lifetimes of different massive star evolutionary phases, which manifests as a change in the ratios of blue-to-red supergiants (B/R) and Wolf–Rayet stars to RSGs (W-R/RSG) in a given population. Van den Bergh (1973) first noted that the B/R in nearby galaxies decreased with the galaxies' absolute magnitude, and suggested that this may be due to a corresponding decrease in chemical abundances (following the luminosity–metallicity relation for star-forming galaxies). Humphreys & Davidson (1979) noted that the B/R in a galaxy or cluster may be representative of the time spent in these respective evolutionary stages; Maeder et al. (1980) expanded upon this further, suggesting that the decrease in B/R at lower metallicities could be due to lower mass-loss rates, which in turn would lead to longer RSG lifetimes and a larger RSG population relative to their blue counterparts.

Maeder et al. (1980) also proposed that this phenomenon should have the same effect on the W-R/RSG ratio. This was observationally confirmed by Massey (2002), finding a factor of over 100 difference in the W-R/RSG ratio over a spread of ~0.9 dex in metallicity (from the SMC to M31, adopting $\log(O/H)+12 = 9.0$ for M31 from Zaritsky et al. 1994; see also Massey 2003 and Figure 6.3). Evolutionary models for massive stars have not yet fully reproduced the change in these ratios, but new treatments of enhanced mass loss, rotation, and binary evolution all show improved agreement with the overall evolution of massive star population ratios (e.g., Meynet & Maeder 2005; Eldridge et al. 2008). It is also important to note that

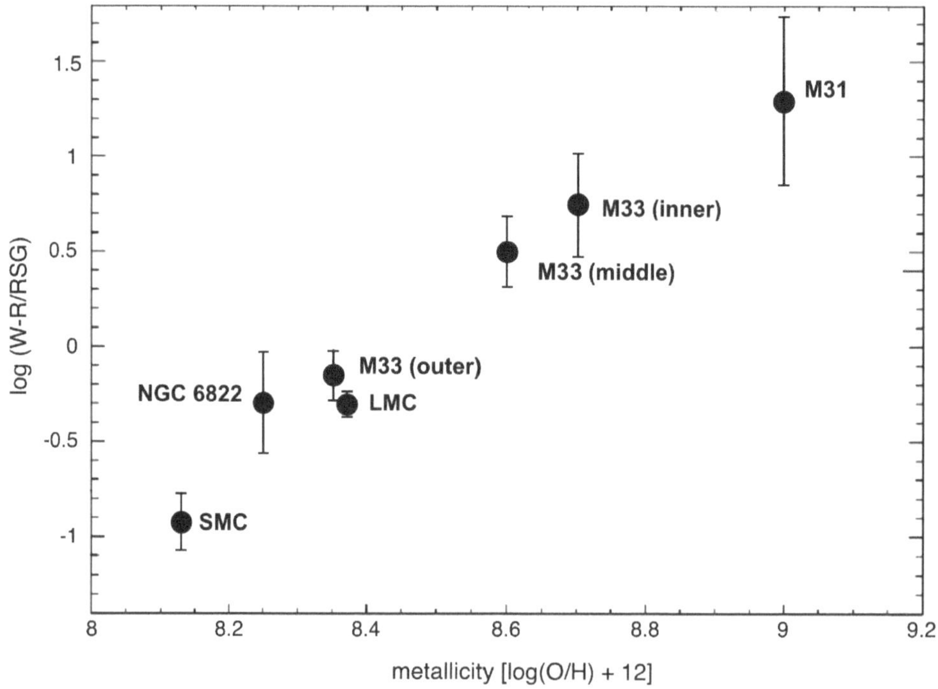

Figure 6.3. Reproduced with permission from Massey (2003; © ESO.); the evolution of the W-R/RSG ratio as a function of metallicity for several galaxies in the Local Group.

this is a somewhat oversimplified picture of the complex massive star populations present in these galaxies, assuming coevality and a uniformity in both completeness and contamination effects across the blue supergiant, RSG, and Wolf–Rayet populations.

Perhaps the starkest metallicity effect in RSG populations is the shift in their spectral types and effective temperatures. In their study of Milky Way and Magellanic Cloud RSGs, Elias et al. (1985) noted a shift in the stars' spectral types, with the average spectral type shifting earlier at lower metallicities. This has since been confirmed in other studies of Magellanic Cloud RSGs (e.g., Massey & Olsen 2003; Dorda et al. 2016), and Levesque & Massey (2012) demonstrated that the effect extends to even lower metallicities (see Figure 6.4, left). It may seem unsurprising that RSG spectral types shift with metallicity—one can imagine the same effective temperature corresponding to weaker TiO bands (and hence earlier spectral types; see Section 3.1) at lower metallicity purely as a consequence of lower abundances, an effect illustrated by the effective temperature scales presented in Levesque et al. (2006; see Figure 6.4, right). However, the impact of metallicity on these spectral type histograms is not simply a spectroscopic effect.

As described in Section 3.2, the minimum effective temperature of an RSG is dictated by the Hayashi limit (Hayashi & Hoshi 1961), which is metallicity-dependent and shifts to warmer temperatures at lower metallicities. Because of this effect, the average *effective temperature*—not merely the average spectral type—

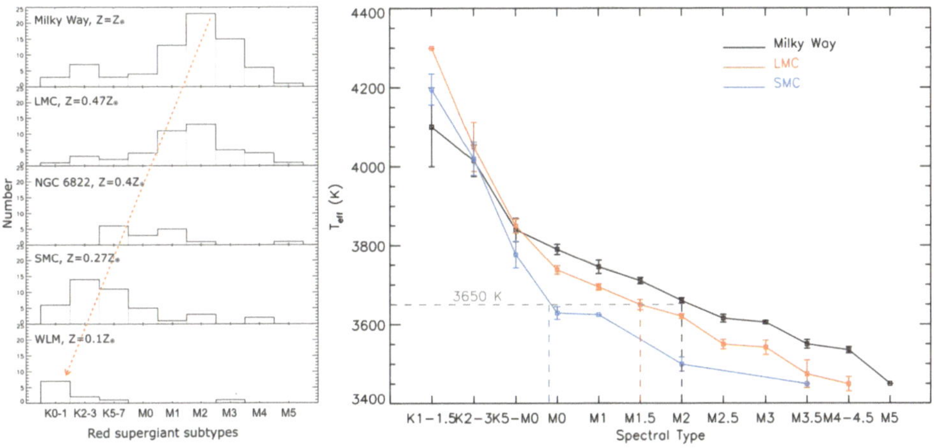

Figure 6.4. Left: reproduced from Levesque & Massey (2012); the observed distribution of RSG spectral subtypes in five different Local Group galaxies, plotted from top to bottom in order of decreasing metallicity. Note the overall progression of the average spectral type toward earlier types at lower metallicities, as well as the handful of late-type outliers in the sub-solar galaxies. Right: reproduced with permission from Levesque (2010); effective temperature scales for RSGs in the Milky Way (black), LMC (red), and SMC (blue). A dashed gray line of constant effective temperature is drawn at 3650 K, and we can see that this same effective temperature would correspond to a spectral type of M2 in the Milky Way, M1.5 in the LMC, and M0 in the SMC, a result of abundance effects on the strengths of the TiO lines.

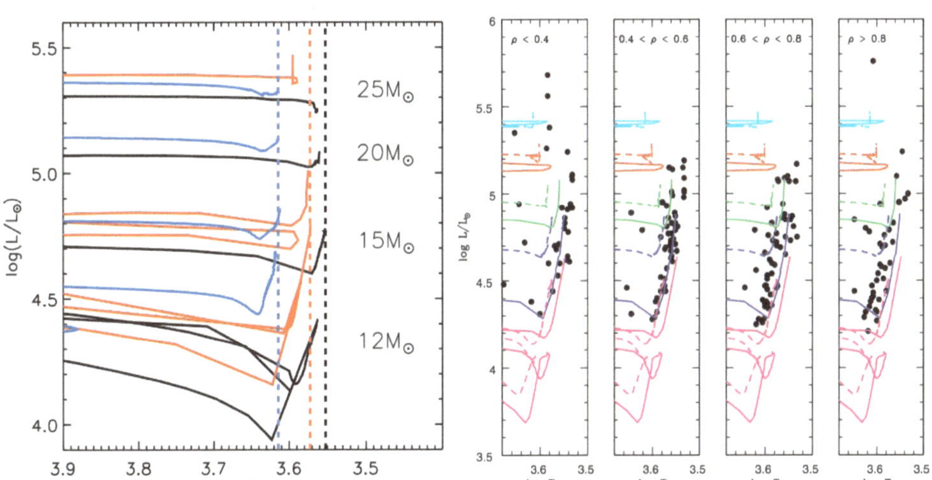

Figure 6.5. Left: reproduced from Levesque et al. (2006); Geneva evolutionary tracks (solid lines) at Milky Way (black; Meynet & Maeder 2005), LMC (red; Schaerer et al. 1993), and SMC (blue; Maeder & Meynet 2001) metallicities, showing the theoretical progression of the Hayashi limit (dashed line) to warmer effective temperatures at lower metallicities. Right: reproduced from Drout et al. (2012); an observational demonstration of the metallicity-dependent Hayashi limit, showing a shift in the effective temperatures of RSGs observed in different regions of M33; a progression from the innermost regions ($\rho < 0.4$) to the outermost regions ($\rho > 0.8$) corresponds to a progression to lower metallicities. Evolutionary tracks are from Ekström et al. (2012).

of an RSG population shifts as a function of metallicity. This effect is predicted by stellar evolutionary tracks (see Figure 6.5, left) and has been demonstrated observationally by Drout et al. (2012) in their study of the RSG population in M33 (see Figure 6.5, right); a progression from the high-metallicity inner regions of M33 to the low-metallicity outskirts corresponds to a clear shift in the position of RSGs on the H-R diagram. This metallicity effect poses an interesting observational challenge when identifying low-metallicity RSG populations photometrically, as one would expect the average color of the RSG population to shift to the blue; however, it is unclear how strong this effect is and in what photometric regime it will be most noticeable (e.g., Chun et al. 2017).

Despite this supposed limit we see RSGs, particularly in metal-poor galaxies, that lie on the "wrong" side of the Hayashi limit in the region of the H-R diagram where stars should not be able to maintain hydrostatic equilibrium (see Figure 6.4, left). With limited data it is unclear whether these stars are in fact in hydrostatic equilibrium (and simply exhibiting an observational effect that is leading to a misinterpretation of their effective temperatures) or truly challenging evolutionary theory with unexpected cold temperatures. One common factor is that almost all of these stars are among the sample of RSGs that display significant spectroscopic variability, which will be discussed in detail in the next chapter.

References

Alcolea, J., Bujarrabal, V., & Gomez-Gonzalez, J. 1990, A&A, **231, 431**

Blaauw, A. 1961, BAN, **15, 265**

Bonanos, A. Z., Lennon D. J., Köhlinger, F., et al. 2010, AJ, **140, 416**

Bonanos, A. Z., Massa D. L., Sewilo M., et al. 2009, AJ, **138, 1003**

Bresolin, F., Pietrzynski G., Urbaneja M. A., et al. 2006, ApJ, **648, 1007**

Britavskiy, N. E., Bonanos A. Z., Mehner A., et al. 2014, A&A, **562, 75**

Britavskiy, N. E., Bonanos A. Z., Mehner A., Boyer M. L., & McQuinn, K. B. W. 2015, A&A, **584, 33**

Caron, G., Moffat, A. F. J., St-Louis, N., Wade G.A., & Lester, J. B. 2003, AJ, **126, 1415**

Clark, J. S., Negueruela, I., Davies B., et al. 2009, A&A, **498, 109**

Chun, S.-H., Sohn, Y.-J., Asplund M., Casagrande, L. et al. 2017, MNRAS, **467, 102**

Cox, N. L. J., Kerschbaum, F., van Marle, A.-J., et al. 2012, A&A, **537, 35**

Davies, B., Figer, D. F., Kudritzki, R.-P., et al. 2007, ApJ, **671, 781**

Davies, B., Kudritzki, R.-P., Plez, B., et al. 2013, ApJ, **767, 3**

Davies, B., Kudritzki, R.-P., Gazak, Z., et al. 2015, ApJ, **806, 21**

Dorda, R., Negueruela, I., González-Fernández, C., & Tabernero, H. M. 2016, A&A, **592, 16**

Dorn-Wallenstein, T., Levesque, E. M., & Ruan, J. J. 2017, ApJ, **850, 86**

Drout, M. R., Massey, P., Meynet, G., Tokarz, S., & Caldwell, N. 2009, ApJ, **703, 441**

Drout, M. R., Massey, P., & Meynet, G. 2012, ApJ, **750, 97**

Ekström, S., Georgy, C., Eggenberger, P., et al. 2012, A&A, **537, 146**

Eldridge, J. J., Izzard, R. G., & Tout, C. A. 2008, MNRAS, **384, 1109**

Elias, J. H., Frogel, J. A., & Humphreys, R. M. 1985, ApJS, **57, 91**

Evans, K. A., & Massey, P. 2015, AJ, **150, 149**
Figer, D. F., MacKenty, J. W., Robberto, M., et al. 2006, ApJ, **643, 1166**
Fujii, M. S., & Portegies Zwart, S. 2011, *Sci*, **334, 1380**
Garmany, C. D., & Stencel, R. E. 1992, A&AS, **94, 221**
Garwood, R., & Jones, T. J. 1987, PASP, **99, 453**
Gazak, J. Z., Davies, B., Kudritzki, R., Bergemann, M., & Plez, B. 2014, ApJ, **788, 58**
Gazak, J. Z., Kudritzki, R., Evans, C., et al. 2015, ApJ, **805, 182**
Gehrz, R., 1989, in Proc. IAU Symp. 135, Interstellar Dust, ed. L. J. Allamandola & A. G. G. M. Tielens (Dordrecht: Kluwer Academic), 445
Gies, D. R., & Bolton, C. T. 1986, ApJS, **61, 419**
Gonález-Fernández, C., Dorda, R., Negueruela, I., & Marco, A. 2015, A&A, **578, 3**
Gvaramadze, V. V., Menten, K. M., Kniazev, A. Y., et al. 2014, MNRAS, **437, 843**
Hayashi, C., & Hoshi, R. 1961, PASJ, **13, 442**
Humphreys, R. M. 1970, AJ, **75, 602**
Humphreys, R. M., & Davidson, K. 1979, ApJ, **232, 409**
Humphreys, R. M., & McElroy, D. B. 1984, ApJ, **284, 565**
Humphreys, R. M., & Sandage, A. 1980, ApJS, **44, 319**
Lardo, C., Davies, B., Kudritzki, R.-P., et al. 2015, ApJ, **812, 160**
Lee, H., Grebel, E. K., & Hodge, P. W. 2003, A&A, **401, 141**
Lee, H., Skillman, E. D., & Venn, K. A. 2005, ApJ, **620, 223**
Levesque, E. M., Massey, P., Olsen, K. A. G., et al. 2005, ApJ, **628, 973**
Levesque, E. M., Massey, P., Olsen, K. A. G., et al. 2006, ApJ, **645, 1102**
Levesque, E. M. 2010, NewAR, **54, 1**
Levesque, E. M., & Massey, P. 2012, AJ, **144, 2**
Mackey, J., Mohamed, S., Neilson, H. R., et al. 2012, ApJL, **751, L10**
Maeder, A., Lequeux, J., & Azzopardi, M. 1980, A&A, **90, L17**
Maeder, A., & Meynet, G. 2001, A&A, **373, 555**
Marco, A., & Negueruela, I. 2013, A&A, **552, 92**
Massey, P. 1998, ApJ, **501, 153**
Massey, P. 2002, ApJS, **141, 81**
Massey, P. 2003, ARA&A, **41, 15**
Massey, P., & Olsen, K. A. G. 2003, AJ, **126, 2867**
Massey, P., Olsen, K. A. G., Hodge, Paul W., et al. 2006, AJ, **131, 2478**
Massey, P., Olsen, K. A. G., Hodge, Paul W., et al. 2007, AJ, **133, 2393**
Massey, P., Silva, D. R., Levesque, E. M., et al. 2009, ApJ, **703, 420**
Massey, P., & Evans, K. A. 2016, ApJ, **826, 224**
Messineo, M., Menten, K. M., Churchwell, E., & Habing, H. 2012, A&A, **537, 10**
Messineo, M., Menten, K. M., Figer, D. F., et al. 2014a, A&A, **569, 20**
Messineo, M., Zhu, Q., Ivanov, V, D., et al. 2014b, A&A, **571, 43**
Messineo, M., Zhu, Q., Menten, K. M., et al. 2016, ApJL, **822, L5**
Meynet, G., & Maeder, A. 2005, A&A, **429, 581**
Negueruela, I., & Schurch, M. P. E. 2007, A&A, **461, 631**
Negueruela, I., Gonz´lez-Fern´ndez, C., Marco, A., Clark, J. S., & Martínez-Núñez, S. 2010, A&A, **513, 74**

Negueruela, I., Gonz´lez-Fern´ndez, C., Marco, A., Clark, J. S., & Martínez-Núñez, S. 2011, A&A, **528, 59**

Neugent, K. F., Massey, P., Skiff, B., & Meynet, G. 2012, ApJ, **759, 177**

Noriega-Crespo, A., van Buren, D., Cao, Y., & Dgani, R. 1997, AJ, **114, 837**

Oh, S., Kroupa, P., & Pflamm-Altenburg, J. 2015, ApJ, **805, 92**

Patrick, L. R., Evans, C. J., Davies, B., et al. 2015, ApJ, **803, 14**

Patrick, L. R., Evans, C. J., Davies, B., et al. 2016, MNRAS, **458, 3968**

Patrick, L. R., Evans, C. J., Davies, B., et al. 2017, MNRAS, **468, 492**

Perryman, M. A. C., de Boer, K. S., Gilmore, G., et al. 2001, A&A, **369, 339**

Poelarends, A. J. T., Herwig, F., Langer, N., & Heger, A. 2008, ApJ, **675, 614**

Ratnatunga, K. U., & Bahcall, J. N. 1985, ApJS, **59, 63**

Sanders, N. E., Caldwell, N., McDowell, J., & Harding, P. 2012, ApJ, **758, 133**

Schaerer, D., Meynet, G., Maeder, A., & Schaller, G. 1993, A&AS, **98, 523**

Siess, L. 2006, A&A, **448, 717**

Slesnick, C. L., Hillenbrand, L. A., & Massey, P. 2002, ApJ, **576, 880**

van den Bergh, S. 1973, ApJ, **183, L123**

Verheyen, L., Messineo, M., & Menten, K. M. 2012, A&A, **541, 36**

Worley, C. C., Irwin, M. J., Tout, C. A., et al. 2016, MNRAS, **459, L31**

Zaritsky, D., Kennicutt, R. C., Jr., & Huchra, J. P. 1994, ApJ, **420, 87**

Chapter 7

Variability in Red Supergiants

7.1 Photometric Variability

Most RSGs display some degree of photometric variability in the optical regime. Changes in V magnitudes on the order of ~ 1 mag are typical, and are seen in RSG populations across a range of metallicities (the typical RSG is much less variable in the near-IR; with $\Delta K \sim 0.25$ mag; e.g., Wood et al. 1983; Josselin et al. 2000; Kiss et al. 2006; Levesque et al. 2007; Massey et al. 2009; Szczygieł et al. 2010). However, despite the ubiquity of this phenomenon, translating observed photometric variations into physical phenomena remains complex.

Observed variability in the optical regime can correspond to a broad range of underlying causes. RSGs are expected to be radial pulsators with luminosity-dependent pulsational periods (e.g., Stothers 1969, 1972; Heger et al. 1997; Guo & Li 2002). As described in Chapter 2, the observed photosphere of an RSG is dominated by the tops of a small number of very large convective cells. At this scale, convection-driven surface effects such as hot spots, shocks (generated from supersonic convection speeds), and cell turnover can all manifest as changes in brightness. Mass loss should also play a role in the optical variability of RSGs; episodic or variable mass loss events could produce observable signatures, and variations in the composition and distribution of the circumstellar dust produced by mass loss can impact the amount of extinction or veiling (see Chapter 4). Finally, effects from a binary companion (including eclipses, mass transfer, and wind interactions; see Chapter 5) can appear as photometric variations.

Disentangling these effects is a daunting (and sometimes impossible) task. The most recent and comprehensive study of variability in RSGs was carried out by Yang & Jiang (2011, 2012), examining the lightcurves of 169 RSGs in the LMC and 112 in the SMC. Of these, 58 LMC stars and 55 SMC stars are classified as irregular variables, with lightcurves that are too complex to be fit with any period (see Figure 7.1, top panel). The remaining stars are classified as "semi-regular" variables (SRc), with this variability further split into two subclasses: short period ($\lesssim$ several hundred

doi:10.1088/978-0-7503-1329-2ch7

days; see Figure 7.1, bottom panel) and "long secondary period" (LSP, $\gtrsim$1000 days; see Figure 7.1, center). LSPs are common in RSGs, while short period variations are less prevalent. A subset of RSGs shows evidence of both in their lightcurves, including Betelgeuse (Stothers 2010), μ Cephei (Kiss et al. 2006), and the Galactic Center RSG GCIRS 7 (Paumard et al. 2014).

There is both theoretical and observational evidence for a period–luminosity (P-L) relation in the pulsating SRc RSGs, particularly those with short period

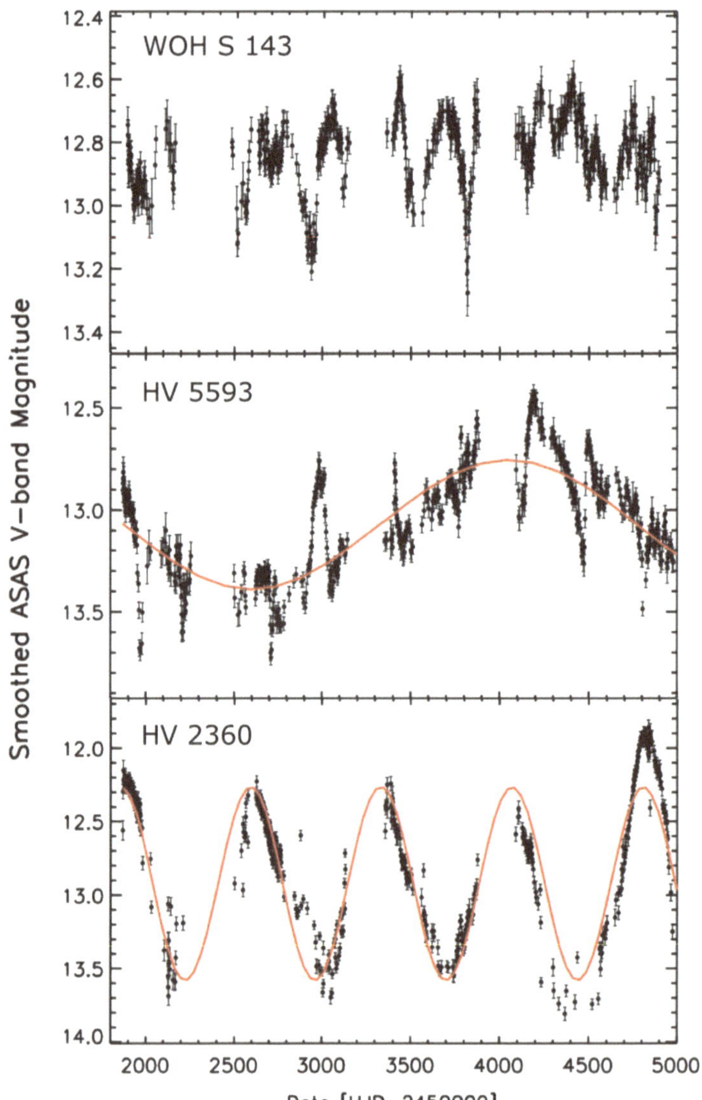

Figure 7.1. Reproduced from Yang & Jiang (2011); All-Sky Automated Survey lightcurves of three LMC RSGs showing examples of the different types of variability: irregular (WOH S 143, top panel), LSP (HV 5593, center panel), and short-period SRc (HV 2360, bottom panel). The red lines represent fitted curves.

variability. Heger et al. (1997) modeled pulsational properties of RSG models between 10–20$M_\odot$ and found that the period of the fundamental mode increased with the luminosity-to-mass ratio. Guo & Li (2002) predicted P-L relations for the first three radial pulsation modes in 15–30$M_\odot$ RSGs, and also predicted a metallicity dependence similar to that seen in AGB stars (with decreasing L and increasing P at higher metallicities). The predicted periods for these range from ~150–4000 days, a result matched by observations (Kiss et al. 2006; Yang & Jiang 2011, 2012). Using the subset of Magellanic Cloud RSGs with distinct short periods—47 in the LMC and 21 in the SMC—Yang & Jiang (2012) found observational evidence of a P-L relation that differed from that seen in Milky Way RSGs.

It is clear that radial pulsations and an associated P-L relation are present in at least some RSGs, but their universality is a matter of debate. We have not yet determined whether radial pulsation is restricted to a certain subset of RSGs, or whether this is a common feature throughout the RSG population that is only detectable in stars without additional sources of variability. The P-L relation has also led to the suggestion that RSGs might be useful as standard candles (e.g., Glass 1979); however, the potential metallicity dependence, the abundance of irregular variables, and the additional sources of long-period variability present significant complications for this prospect.

While pulsation models do predict periods of several thousand days, this is only one potential explanation for the LSPs, which are seen in the majority of SRc RSGs studied by Yang & Jiang (2011, 2012). Surface effects from convection are also expected to play a dominant role in the observed variability of RSGs, although the periodicity of these effects is not well established (e.g., Josselin & Plez 2007, Chiavassa et al. 2009, 2011). Convection processes at the surfaces of RSGs can produce variable hot spots (Baron et al. 2014), observed in surface imaging of Betelgeuse and other RSGs (e.g., Haubois et al. 2009; Chiavassa et al. 2010, Montargès et al. 2016; O'Gorman et al. 2017). Simple convective turnover in giant (~400–500$R_\odot$) convective cells can also contribute to brightness variations on a timescale of years (e.g., Stothers 2010; Chiavassa et al. 2009). Finally, shocks in the atmospheres of RSGs (see Chiavassa & Freytag 2015 and Chapter 2) can excite sporadic emission features, which can temporarily increase the observed flux in specific wavelength regimes.

The potential roles of mass loss and dust production in RSGs are still unclear. As described in Chapter 4, there are multiple mechanisms that can contribute to mass loss and dust production in RSGs, and some of these same mechanisms can also contribute to brightness variations (such as the radial pulsations described above). Sporadic or variable mass loss is predicted in RSGs based on observations of supernovae; Yoon & Cantiello (2010) predict dramatic and swift changes in the wind-driven mass-loss rate of RSGs, and Yaron et al. (2017) find evidence that the RSG progenitor of SN 2013fs ejected a dense shell of material at a highly elevated rate of mass loss just prior to its explosion. Pre-supernova instabilities in RSGs could potentially manifest as pre-explosion brightness changes (although there is not yet any observational evidence that they experience the same massive pre-explosion eruptions as the presumed LBV progenitors of Type IIn supernovae; e.g., Ofek et al. 2014; Levesque et al. 2014a; for more discussion see Chapter 8).

Variable mass loss in RSGs can also produce variations in the amount and structure of their circumstellar dust. Depending on the distribution and grain size of the dust (see Chapter 4), this can appear observationally as a temporary change in the color of the star or as an increase in the veiling of the star's spectrum as a whole. Dust-enshrouded OH/IR stars show evidence of significantly greater photometric variability in the optical as compared to the general RSG population (see Section 4.3), and other variable RSGs show changes in their visual extinction on a timescale of months (see Section 7.2).

All of these phenomena can contribute to changes in the apparent magnitude of an RSG and the resulting periodic or stochastic fluctuations in its lightcurve. However, without multi-wavelength high-cadence monitoring of variable RSGs, it is difficult to determine whether these variations are a result of changes in the star's observed brightness, color, bolometric luminosity, or some combination of the three. Multiple observations of these variable stars' spectra at key points in their lightcurve can offer another means of connecting their variability with a physical origin. This also allows us to explore the much rarer phenomenon of spectroscopic variability in RSGs.

7.2 Spectroscopic Variability

Minor spectroscopic variability in RSGs is not unsurprising when considering the physics of their outer layers: extended atmospheres, optical depth effects near the Hayashi limit, and effects from the same phenomena listed above (e.g., convection, pulsation, and mass loss) can all contribute to small variations in the widths and strengths of spectral features. Determining the spectral type of an RSG is also challenging and typically qualitative rather than quantitative, based primarily on the continuum shape for early K-type stars and the comparative strengths of the TiO bands in late K- and M-type RSGs (see Section 3.1). Differences in spectral type on the order of a subtype or less can sometimes be attributed to errors or inconsistencies in the methods used for spectral typing (see Levesque et al. 2007).

White & Wing (1978) examined spectral variations in 128 Milky Way RSGs and identified 28 stars in the sample whose spectral type changed by more than half a subtype; with this criteria, it is likely that a number of these apparent variations can be attributed to normal inconsistencies in classification. More recently, Dorda et al. (2016) obtained spectroscopy of ~200 late-type (G, K, and M) supergiants in the Magellanic Clouds (108 in the SMC and 79 in the LMC) at multiple epochs (spanning one to three years), and found that 33% of the LMC stars and as many as 84% of the SMC stars showed spectral type variations of one subtype or more; they also concluded that the mean variation in spectral type is larger in the SMC sample. However, it is worth noting that Dorda et al. (2016) do not use the standard means of measuring spectral types in cool supergiants - they determine their spectral types quantitatively using a set of three equivalent width indices (combining 5 Ti I lines, 12 Fe I lines, and the Ca II triplet), and note that this method breaks down as TiO bands become stronger and start to blanket out the continuum, limiting its utility for RSGs with spectral types of K5 and later.

True spectroscopic variability—changes of a spectral type or more that correspond to clear changes in the star's spectrum and presumed physical properties—has only been documented in a small number of RSGs. The most dramatic example is the SMC RSG HV 11423, described by Massey et al. (2007). Originally observed in December 2004, the star was assigned a spectral type of K0-1 and an effective temperature of 4300 K. It was then observed again in December 2005 and found to be much later and cooler, with an M4 spectral type and a 3500 K effective temperature. A third spectrum in the blue was taken in September 2006 and agreed well with the spectral energy distribution of the K0-1 2004 spectrum. A fourth spectrum in August 2010 showed the star in an interim stage of M0 I and 3650 K, while a fifth in September 2013 corresponded to a slightly later and cooler spectrum with M1.5 and 3575 K (see Figure 7.2). In addition to these extreme and rapid variations in spectral type and apparent effective temperature, HV 11423 also displays abnormally high variability in V, well in excess of the ~1 mag variations typical of RSGs and discussed above. It also showed considerable variations in its bolometric luminosity and circumstellar reddening, appearing brighter, dustier, and more luminous in its warm early-type state.

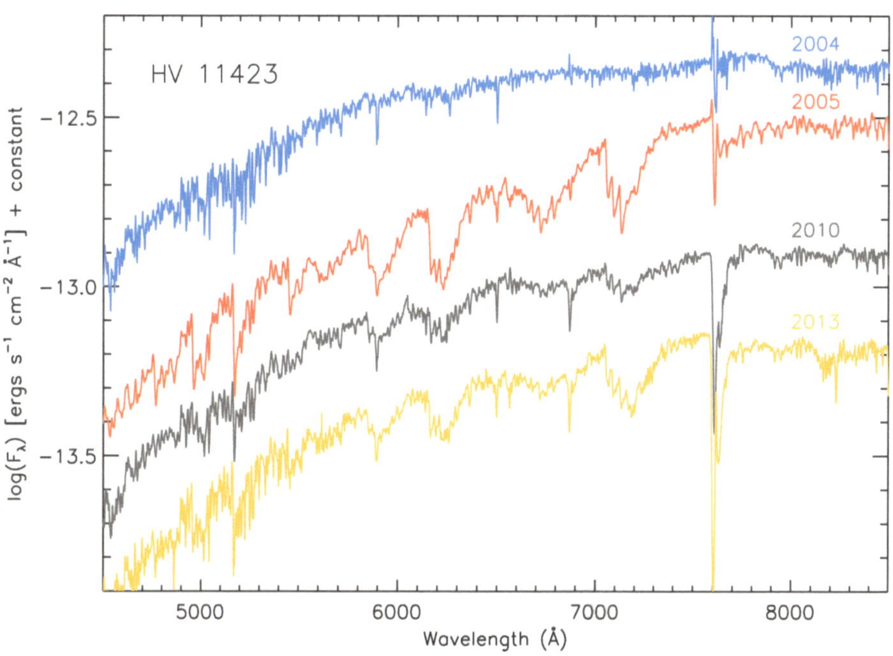

Figure 7.2. Spectrophotometry of HV 11423 over a nine-year period, illustrating the star's variations in spectral type and (presumed) effective temperature. The observations were taken in Dec 2004, Dec 2005 (Massey et al. 2007), Aug 2010 (Levesque & Massey 2012), and Sep 2013 (K. Evans, 2017, private communication), and correspond to spectral types (and effective temperatures) of K0 (~4300 K), M4 (3500 K), M0 (3625 K), and M1.5 (3575 K), respectively.

HV 11423 was originally identified as a single unusual star, but Levesque et al. (2007) later found three more RSGs—SMC 055188, SMC 046662, and LMC 170452—that fit into the same variability template. All of these stars exhibit unusual variability in their optical spectra, *V* magnitudes, bolometric luminosity, and circumstellar reddening on a timescale of months, and the stars are brighter, dustier, and more luminous when they display their earliest spectral types. In their coolest states, each of these stars also falls well to the right of the Hayashi limit on the H-R diagram, occupying (perhaps unsurprisingly) a region of the diagram where stars are not expected to be in hydrostatic equilibrium. Levesque & Massey (2012) identified two additional RSGs—J194453.46-144552.6 in NGC 6822 and J000158.14-152332.2 in WLM—that also have atypically late spectral types for their host galaxies, suggesting that their effective temperatures are also cooler than the Hayashi limit. Interestingly, WOH G64, the dust-enshrouded LMC OH/IR RSG (see Section 4.3), does *not* show any evidence of spectral variability, despite having an extremely late spectral type (M5) for the LMC. Fellow OH/IR star VY CMa is similarly constant in spectral type (Wallerstein 1958; Wallerstein & Gonzalez 2001).

Several other stars have been proposed as candidate members of this same class based on observed changes in spectral type. The RSG J194449.96-144333.5 in NGC 6822 was classified as a K5 star by Levesque & Massey (2012) and a M2.5 star by Massey (1998), suggesting that it may be yet another member of this spectroscopically variable class. Bonanos et al. (2009, 2010) further identify RSGs in the Magellanic Clouds that show disparities between their colors and temperatures relative to the general population, suggesting that these may also be spectroscopic variables (one, SMC 049990, was classified as a K5 by Massey & Olsen (2003) but a G2 by Evans et al. 2004). Lightcurves of the spectroscopically variable LMC and SMC candidates suggest that they would likely be classified as SRc variables with relatively short periods (~1–2 years). Wasatonic et al. (2015) identified the RSG TV Gem in the Milky Way as a potential spectroscopic variable. It has both a short-term period (~1.2 years) and LSP (~8.59 years), and it is quoted as ranging from M0 to "M4.1" in spectral type over the course of its LSP (following the Wing 1992 method of assigning spectral types based on indices computed from narrow imaging filters centered on the 7054Å TiO band and surrounding continuum). Ironically, this star was classified as a spectral standard by Morgan & Keenan (1973). It was also assigned consistent spectral types of M0-1 in Oct 2001 (Massey & Olsen 2003) and M1 in May 2004 (Levesque et al. 2005); however, its lightcurve and suggested TiO band variations still make it an intriguing target.

While this variability has not yet been fully explained, the changes in these stars are characteristic of sporadic mass loss leading to changes in the opacity of the photosphere and the quantity and distribution of circumstellar dust (consistent with the circumstellar dust shells previously described by Danchi et al. 1994 and others; see Section 4.1). Low metallicity also appears to play a role (a conclusion also suggested by Dorda et al. 2016), as the variables and candidates are heavily concentrated in sub-solar metallicity galaxies.

Some RSGs have constant spectral types based on their TiO bands but exhibit variable *emission* line spectra. This is a signature of the VV Cep binaries, which show variable Balmer and [Fe II] emission features in the optical and Fe II emission in the UV (see Section 5.2). The OH/IR star VY CMa sporadically displays emission features in its spectrum on a timescale of months; WOH G64 shares these emission features and also displays a nebular emission spectrum, though the variable nature of these features has not been established (Levesque et al. 2009; see Section 4.3). Other RSGs show evidence of emission lines in their spectra, either from the star itself or from an ionized circumstellar nebula, but so far there has been no comprehensive study (and certainly no time-resolved data) on emission lines in RSG spectra. The TŻO candidate HV 2112 displays dramatic photometric, spectroscopic, *and* emission line variability (e.g., Levesque et al. 2014b, see Section 5.3.3).

References

Baron, F., Monnier, J. D., Kiss, L. L., et al. 2014, ApJ, **785, 46**

Bonanos, A. Z., Massa, D. L., Sewilo, M., et al. 2009, AJ, **138, 1003**

Bonanos, A. Z., Lennon, D. J., Köhlinger, F., et al. 2010, AJ, **140, 416**

Chiavassa, A., Plez, B., Josselin, E., & Freytag, B. 2009, A&A, **506, 1351**

Chiavassa, A., Haubois, X., Young, J. S., et al. 2010, A&A, **515, 12**

Chiavassa, A., Pasquato, E., Jorissen, A., et al. 2011, A&A, **528, 120**

Chiavassa, A., & Freytag, B., 2015, in ASP Conf. Ser. 497, Why Galaxies Care about AGB Stars III: A Closer Look in Space and Time, ed. F. Kerschbaum, R. F. Wing, & J. Hron (San Francisco, CA: ASP), 11

Danchi, W. C., Bester, M., Degiacomi, C. G., Greenhill, L. J., & Townes, C. H. 1994, AJ, **107, 1469**

Dorda, R., Negueruela, I., González-Fernández, C., & Tabernero, H. M. 2016, A&A, **592, 16**

Evans, C. J., Howarth, I. D., Irwin, M. J., Burnley, A. W., & Harries, T. J. 2004, MNRAS, **353, 601**

Glass, I. S. 1979, MNRAS, **186, 317**

Guo, J. H., & Li, Y. 2002, ApJ, **565, 559**

Haubois, X., Perrin, G., Lacour, S., et al. 2009, A&A, **508, 923**

Heger, A., Jeannin, L., Langer, N., & Baraffe, I. 1997, A&A, **327, 224**

Josselin, E., Blommaert, J. A. D. L., Groenewegen, M. A. T., Omont, A., & Li, F. L. 2000, A&A, **357, 225**

Josselin, E., & Plez, B. 2007, A&A, **469, 671**

Kiss, L. L., Szabó, Gy. M., Bedding, T. R., et al. 2006, MNRAS, **372, 1721**

Levesque, E. M., Massey, P., Olsen, K. A. G., et al. 2005, ApJ, **628, 973**

Levesque, E. M., Massey, P., Olsen, K. A. G., & Plez, B. 2007, ApJ, **667, 202**

Levesque, E. M., Massey, P., Plez, B., & Olsen, K. A. G. 2009, AJ, **137, 4744**

Levesque, E. M., & Massey, P. 2012, AJ, **144, 2**

Levesque, E. M., Stringfellow, G. S., Ginsburg, A. G., Bally, J., & Keeney, B. A. 2014a, AJ, **147, 23**

Levesque, E. M., Massey, P., Zytkow, A. N., & Morrell, N. 2014b, MNRAS, **443, L94**

Massey, P. 1998, ApJ, **501, 153**

Massey, P., Levesque, E. M., Olsen, K. A. G., Plez, B., & Skiff, B. A. 2007, ApJ, **660, 301**

Massey, P., & Olsen, K. A. G. 2003, AJ, **126, 2867**

Massey, P., Silva, D. R., Levesque, E. M., et al. 2009, ApJ, **703, 420**

Morgan, W. W., & Keenan, P. C. 1973, ARA&A, **11, 29**

Montargès, M., Kervella, P., Perrin, G., et al. 2016, A&A, **588, 130**
Ofek, E. O., Sullivan, M., Shaviv, N. J., et al. 2014, ApJ, **789, 104**
O'Gorman, E., Kervella, P., Harper, G. M., et al. 2017, A&A, **602, L10**
Paumard, T., Pfuhl, O., Martins, F., et al. 2014, A&A, **568, 85**
Stothers, R. 1969, ApJ, **156, 541**
Stothers, R. 1972, A&A, **18, 325**
Stothers, R. B. 2010, ApJ, **725, 1170**
Szczygieł, D. M., Stanek, K. Z., Bonanos, A. Z., et al. 2010, AJ, **140, 14**
Wallerstein, G. 1958, PASP, **70, 479**
Wallerstein, G., & Gonzalez, G. 2001, PASP, **113, 954**
Wasatonic, R. P., Guinan, E. F., Durbin, A. J., et al. 2015, PASP, **127, 1010**
White, N. M., & Wing, R. F. 1978, ApJ, **222, 209**
Wing R. E. 1992, JAVSO, **21, 42**
Wood, P. R., Bessell, M. S., Fox, M. W., et al. 1983, ApJ, **272, 99**
Yang, M., & Jiang, B. W. 2011, ApJ, **727, 53**
Yang, M., & Jiang, B. W. 2012, ApJ, **754, 35**
Yaron, O., Perley, D. A., Gal-Yam, A., et al. 2017, NatPh, **13, 510**
Yoon, S-C., & Cantiello, M., 2010, ApJ, **717, 62**

Chapter 8

Red Supergiants and Supernovae

The mass range of RSGs makes them a critical transitional stage for massive stars reaching the ends of their core-fusion lives and preparing to explode as supernovae (SNe). At lower masses, RSGs are the direct progenitors of core collapse; at higher masses they are a presumed intermediate evolutionary stage before the star undergoes core collapse as a yellow supergiant, blue supergiant, or Wolf–Rayet star. As a result, it is worthwhile to specifically explore RSGs as SN progenitors, considering the properties that we are able to infer from direct progenitor detections and upper limits and the implications of their evolution for the deaths of massive stars.

8.1 RSGs as Direct Progenitors - Type II-P Supernovae

RSGs are the direct progenitors of Type II-P supernovae. Clear hydrogen lines in the supernova spectrum give it its "Type II" designation (as opposed to the hydrogen-free Type I supernovae), while the "P" indicates the presence of a distinctive plateau in the supernova lightcurve that occurs shortly after maximum (setting it apart from the more linear decay of other supernovae; see Figure 8.1). The plateau typically lasts several dozen days and is interpreted as a result of the interaction between the supernova shock wave and the RSG's thick envelope: as the shock wave propagates outward it heats the outer layers of the envelope and ionizes the hydrogen, increasing the opacity. The net effect of this is to temporarily maintain a constant radius for the photosphere even as the ejecta expands; the plateau then ends when the temperature cools sufficiently to allow the hydrogen to recombine (Filippenko 1997).

A small number of Type II-P supernovae have been directly associated with specific RSG progenitors thanks to the existence of pre-explosion imaging of their host environments. About a dozen Type II-P supernovae have been robustly associated with a progenitor star observed in pre-explosion imaging (e.g., Smartt 2015; Maund 2017), despite the challenges associated with this technique. Because pre-explosion images are often not tailored to specifically detect these stars,

doi:10.1088/978-0-7503-1329-2ch8

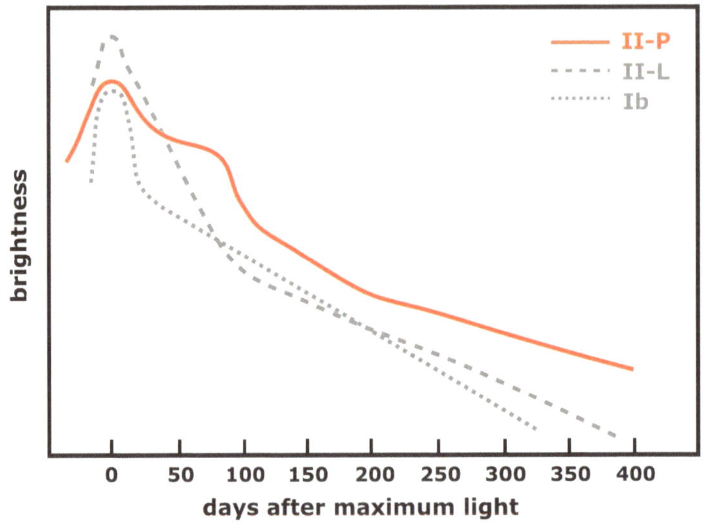

Figure 8.1. Schematic illustration of a typical Type II-P supernova lightcurve (red), compared to the lightcurves of Type II-L (gray dashed) and Type Ib (gray dotted) supernovae. The plateau featured in the II-P lightcurve can extend from ~50–100 days after maximum light.

progenitor detections are often extremely faint and can be complicated by the presence of multiple or blended sources that agree with the SN position to within astrometric errors. Even when detections are statistically significant, translating observations into reliable photometry is very dependent on the specific techniques used for measuring faint sources, which can increase errors (e.g., Maund et al. 2014). Even if pre-explosion images are available in multiple filters, the progenitor is often only detected in one, with the other images offering only upper limits. Some observations initially consistent with a progenitor star have later been revised. To take one example, pre-explosion imaging of the SN 1999ev explosion site in NGC 4274 suggested a 4.8σ detection of an object with HST/WFPC2, with a quoted $m_{\rm F555W} = 24.64 \pm 0.17$; however, Maund et al. (2014) later determined that the source was still present at a comparable brightness in post-explosion imaging, and suggested that the source might instead be a compact stellar cluster (which perhaps hosted the SN). Similar detections for candidate host clusters have been determined for other Type II-P supernovae, including SN 2004dj and SN 2004am (see Smartt 2015). However, in the best cases, the progenitor is detected in multiple filters (or strong upper limits can be applied), and the resulting color information can demonstrate that the star is likely an RSG (e.g., Van Dyk et al. 2012a, 2012b). Follow-up imaging (taken well after the supernova has faded) can also confirm this association by revealing that the progenitor star has disappeared (see Figure 8.2).

Despite the challenges associated with interpreting pre-explosion imaging, these identifications are a compelling potential source of information regarding the physical properties and evolutionary states of the progenitor just before its explosion. As these identifications often occur in relatively nearby galaxies with well-known distances and metallicities, an estimate of the star's effective

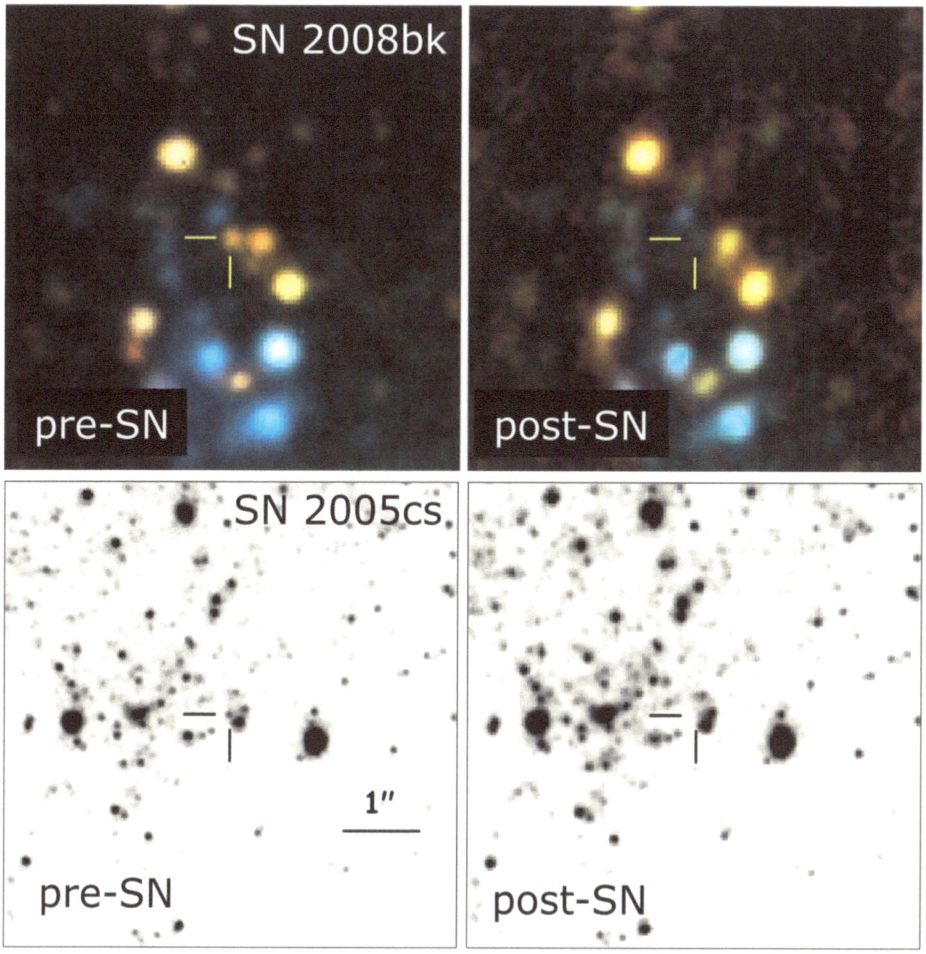

Figure 8.2. The two best cases of RSG SN progenitors detected in pre- and post-explosion imaging. Top panel: adapted from Mattila et al. (2010). The left panel shows pre-explosion VIK_s imaging of the SN 2008bk explosion site (marked with yellow crosshairs) from the VLT, combining V and I band imaging from 16 Sept 2001 with K_s imaging taken on 17 Oct 2005; the right panel shows post-explosion VIK_s imaging of the same site from the NTT, combining V and I band imaging from 16 Sep 2010 with K_s imaging from 29 Oct 2010. Bottom: reproduced from Maund et al. (2014). The left panel shows pre-explosion (0.44 years before discovery) HST ACS/WFC F814W imaging of the SN 2005cs explosion site (marked with crosshairs); the right panel shows HST ACS/WFC F814W imaging of the same site 5.09 years after the discovery of the SN.

temperature and luminosity can allow it to be placed on the H-R diagram, where a comparison to evolutionary tracks can be used to estimate a progenitor mass. However, it is worth carefully considering the potential sources of error present in this method and how these affect the errors and precision of the resulting physical properties inferred for the progenitor.

To begin, estimating an absolute magnitude for a progenitor requires accounting for the effects of dust. Foreground and host reddening can be well-constrained by post-explosion observations, but dust produced by the progenitor itself is likely destroyed in the supernova explosion, rendering it impossible to quantify the amount (and reddening law) of this dust in follow-up observations. As demonstrated in Chapter 4, RSGs produce a substantial amount of circumstellar dust, which can significantly impact their optical photometry. Photometric variability in RSGs is also common, with $\Delta V \sim 1$ mag being typical for normal RSGs (see Chapter 7), introducing an additional source of error. Translating the calculated absolute magnitude to a luminosity then requires applying a bolometric correction, which is substantial and strongly temperature-dependent in the optical due to the cool effective temperatures typical of RSGs (see Chapter 3).

Determining an effective temperature for an RSG based only on imaging data is also difficult. With detections or constraining upper limits in multiple bands, pre-explosion photometry can be compared to model SEDs (ideally, those that include treatments of atmospheric geometry such as the MARCS and PHOENIX models; see Chapter 2) to at least estimate a best-fit temperature; however, any degree of precision is generally limited by ambiguities in the strengths of the TiO bands and the wavelength coverage and throughput of the specific filters. That said, in practice the SED fitting method (or even single-point photometry combined with upper limits from bluer filters), *is* a good means of establishing a rough maximum temperature by establishing the blackbody temperature that can accommodate blue data. A lower limit can also reasonably be estimated by adopting the effective temperature of the Hayashi limit at the metallicity of the host galaxy (while this fails to account for spectroscopically variable RSGs that sometimes appear below the Hayashi limit, such stars seem to be relatively rare, especially at high metallicities; see Chapter 7). These limits thus make it possible to at least bracket the maximum and minimum bolometric corrections that should be applied.

As detailed in Section 3.4, luminosity can function as a proxy for mass when plotting RSGs on the H-R diagram; however, the luminosities of stars with different initial masses overlap considerably in the RSG regime, making a precise mass determination difficult even with very well-constrained stellar luminosities. The overlap is further exacerbated when we consider how different phenomena can impact the RSG and terminal evolution of a massive star. This is best illustrated by comparing different treatments of stellar evolution on a single H-R diagram (Figure 8.3). Differences between non-rotating and rotating models, single and binary models (see also Chapter 5), and even model-specific differences in how mass loss and convection are treated can all dramatically impact the predicted range of luminosities for a given initial mass. Figure 8.3 demonstrates that the combination of luminosity error bars and changes in model treatments can yield an uncertainty of $\gtrsim 8 M_{\odot}$. Finally, stellar evolution during the final years and final core-fusion stages of a massive star's life is still poorly understood, and RSGs could experience significant changes in their physical properties (and associated mass loss and binary interactions) even in the final few years of their lives (e.g., Yoon & Cantiello 2010; Shiode & Quataert 2014).

Finally, while the vast majority of Type II-P SNe do not benefit from the luck of having existing pre-explosion imaging, these events *do* offer their own observables that can sometimes be used to extract estimates of progenitor properties. Supernova models suggest relations between Type II-P explosion energy and upper limits on the

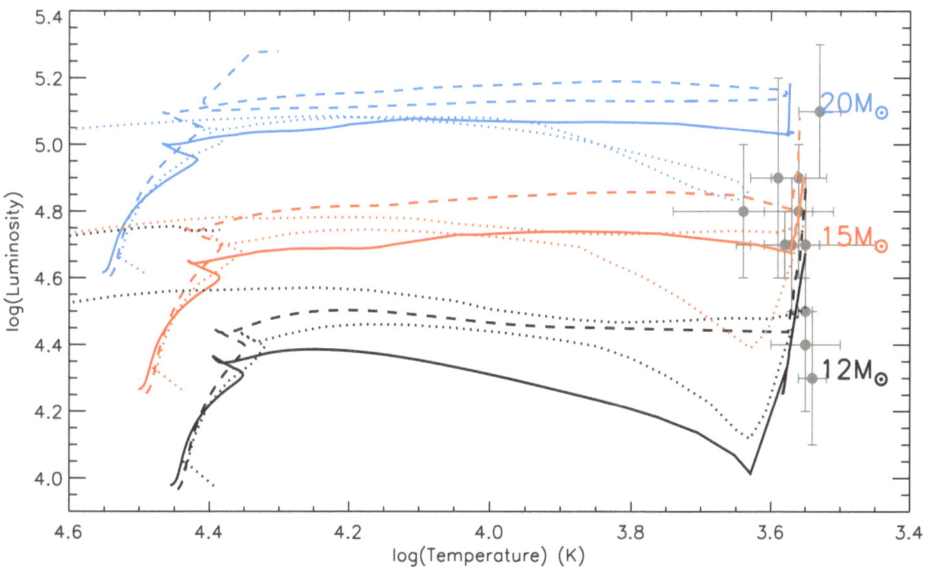

Figure 8.3. Comparison of $12M_{\odot}$ (black), $15M_{\odot}$ (red), and $20M_{\odot}$ (blue) evolutionary tracks at solar metallicity with different treatments of binary and rotation effects. Solid tracks and dashed lines are, respectively, single-star non-rotating and rotating Geneva models from Ekström et al. (2012); dotted tracks are BPASS models (e.g., Eldridge & Stanway 2016) modeling a primary star with the quoted initial mass in a binary, adopting a secondary-to-primary mass ratio of 0.5 and a period of 100 days. There is a substantial overlap at the Hayashi limit between tracks with different initial masses. Note also that the $20M_{\odot}$ rotating single-star model and *all* of the binary models predict a blueward evolution of the star prior to core collapse. The inferred luminosities and temperatures of 11 Type II-P supernova progenitors listed in Smartt (2015) are overplotted in light gray, illustrating the range of initial masses consistent with each data point.

progenitor's initial mass (e.g., Dessart et al. 2010) or between plateau brightness, the temperature evolution of the SN emission, and progenitor radius (e.g., Dessart & Hillier 2011; Rabinak & Waxman 2011). The ^{56}Ni mass inferred for the supernova also appears to correlate with its maximum luminosity (Pejcha & Prieto 2015), which can in turn be translated back to a progenitor radius. Fitting hydrodynamical models to the lightcurves can also be used to estimate ejecta masses (although these are generally much higher than, and in conflict with, those inferred from direct progenitor observations, e.g., Maguire et al. 2010), and late-time spectral modeling can be used to estimate nucleosynthesis yields and thus establish a progenitor mass upper limit (e.g., Jerkstrand et al. 2012). Finally, studies of the close resolved stellar populations around the explosion site can be carried out at any time pre- or post-explosion and can sometimes provide constraints on age, and as a result, approximate maximum or minimum progenitor masses (e.g., Williams et al. 2014; Jennings et al. 2014; Maund 2017). In addition, some post-explosion observations of Type II-P supernovae can offer evidence of pre-explosion mass loss; Yaron et al. (2017) suggest that the progenitor of SN 2013fs experienced a bout of high mass loss shortly before its explosion and ejected a dense shell of mass (see also Section 7.1).

For all of these reasons, high-precision estimates of effective temperature, luminosity, and mass for RSG progenitors based on pre-explosion imaging or SN modeling are not feasible. However, even very approximate constraints on progenitor properties can be invaluable given the rarity of pre-explosion detections and the paucity of information we currently have on how to accurately map progenitor properties to core-collapse explosions. Reasonable initial mass estimates for RSGs must include fairly broad and significant error bars (in light of the strong model-dependence of mass estimates and the significant errors present in luminosity and effective temperature estimates) but these are still able to exclude some mass ranges as II-P progenitors, even in cases where pre-explosion imaging can only give photometric upper limits. One interesting result of these limits has been dubbed the "red supergiant problem."

8.2 The "Red Supergiant Problem"

First described by Smartt et al. (2009), this term refers to the clear lack of high-mass ($\gtrsim 17 M_{\odot}$) RSGs that have been detected as supernova progenitors. This is at odds with the mass range of RSGs, and especially surprising when considering that the observational sample should be skewed toward the most massive and luminous progenitors (as they should be the easiest to detect in pre-explosion imaging).

One possible explanation is that this problem is merely a casualty of large (and frequently underestimated) error bars in mass determinations for RSG progenitors. Adopting a more diverse range of evolutionary tracks for mass estimates and fully accounting for the various sources of error could expand the "measured" mass range of RSG progenitors to the point where this apparent high-mass gap disappears. However, the body of work supporting a surprising lack of high-mass RSGs as SN progenitors is substantial (see Smartt 2015) and argues that alternative explanations should be strongly considered.

Walmswell & Eldridge (2012) suggest that this problem is a consequence of not accounting for extinction effects from the dust produced in RSG winds. A re-estimate of the maximum initial mass for detected Type II-P progenitors suggests $\sim 21 M_{\odot}$, in better agreement with the predicted maximum RSG mass. Similarly, it is possible that many of the most massive SN progenitors are simply too heavily reddened to be detected in optical pre-explosion imaging; the OH/IR stars, for example, are extremely massive but also heavily dust-enshrouded and would be impossible to detect at the distances typical of Type II-P supernovae (see Section 4.3). Beasor & Davies (2016) reach a very similar conclusion to Walmswell & Eldridge (2012) when considering LMC RSGs as SN progenitors, and find that the most luminous (and presumably most massive and evolved) RSGs would in fact be the most susceptible to increased extinction (although it is also worth noting that Kochanek et al. 2012 find that this may translate to a minimal loss in flux as a result of light being scattered back into the line of sight).

The maximum mass of an RSG and the maximum mass of a *terminal* RSG are also two different things. At higher initial masses RSGs evolve leftward on the H-R diagram prior to core collapse, spending some amount of time as yellow supergiants, blue supergiants, or Wolf–Rayet stars before to exploding (e.g., Ekström et al. 2012). The exact mass limit at which this occurs is complex, but it lies in the 20–30$M_\odot$ regime, suggesting that this could explain the lack of explosions seen occurring directly from high-mass RSGs (see Section 8.3 for further discussion).

Type II-P supernovae are also not the only proposed explosive end phase originating from RSGs. Smith et al. (2015) highlighted the interesting case of PTF11iqb, which was initially classified as a Type IIn supernova (characterized by narrow spectral lines associated with the interaction of the supernova shock with dense circumstellar material), but quickly evolved to more closely resemble a Type II-P or II-L (linear decay) supernova. While Type IIn supernovae are generally associated with luminous blue variables (e.g., Ofek et al. 2014; Levesque et al. 2014), this event suggests that some narrow-line supernovae can arise from lower-mass and cooler progenitors that have undergone similar high mass loss near the ends of their lives. However, while this effect could also render them difficult to detect in pre-explosion imaging (due to the increased extinction effects described above), it is worth noting that an RSG-like progenitor has not yet been identified in pre-explosion imaging for a Type IIn or Type II-L supernova.

Finally, another intriguing explanation is the prospect that some more massive RSGs are dying through direct collapse to a black hole, which would lack the rebound explosion of a supernova and lead to the simple disappearance of the RSG with only a very faint or completely absent optical transient. Such a collapse is predicted for a variety of different massive star models, depending on treatments of their interior physics (e.g., Ugliano et al. 2012; Lovegrove & Woosley 2013; Sukhbold & Woosley 2014; Pejcha & Thompson 2015; Sukhbold et al. 2016). Adams et al. (2017) recently identified the first potential observed candidate for an RSG directly collapsing to a black hole. The object N6946-BH1, originally identified by Gerke et al. (2015), is in the nearby galaxy NGC 6946 and had a spectral energy distribution that was broadly consistent with a $\sim$25$M_\odot$ RSG at the host's roughly solar metallicity. After a decade of relatively constant bolometric luminosity, the star underwent a weak outburst in 2009 followed by an abrupt drop in the optical luminosity and a slower fade in the mid-IR. It is now well below its previously constant bolometric luminosity, with an evolution that is consistent with a "failed supernova" direct black hole collapse (rather than a stellar merger or an abrupt change in circumstellar extinction). Finally, gravitational waves have recently been detected from three different pairs of merging black holes with masses ranging from 7.4 ± 2.3 to 36^{+5}_{-4} (Abbott et al. 2016a, 2016b; Abbott et al. 2017); the masses and implied population numbers of these black holes are consistent with the production of failed SNe from stars in the mass range of RSGs (e.g., Belczynski et al. 2016; Woosley 2016).

8.3 RSGs as Progenitor Precursors

While lower-mass RSGs undergo core collapse immediately following the RSG phase, higher-mass RSGs evolve back across the H-R diagram in the later stages of their evolution, undergoing the dramatic mass loss that leads to another brief post-RSG yellow supergiant phase, and, eventually, the Wolf–Rayet phase before core collapse occurs. As a result, while RSGs are not the *direct* progenitors of these stars' supernovae, their evolution during the RSG phase has a substantial impact on the SN progenitors that they ultimately produce.

The exact mass where this occurs is a strong function of parameters such as metallicity, rotation, and binary interaction. Introducing rotation, for example, decreases the maximum mass at which a star is expected to die as an RSG from $20M_\odot$ to $15M_\odot$ (see Figure 8.4). Rotational mixing in these stars leads to longer main-sequence lifetimes (as a result of mixing additional hydrogen into the core), stronger surface enrichments, and an anticipated higher mass-loss rate in the RSG phase. This effectively shortens RSG lifetimes overall and makes a blueward evolution at late ages easier (for more, see Ekström et al. 2012; Levesque et al. 2012). Metallicity has a similar effect; at higher metallicities the maximum mass for a terminal RSG will also drop due to an expected increase in mass-loss rate during the RSG phase (this is seen in the evolution of the WR/RSG ratio with metallicity; see Section 6.4). Ambiguities in this mass limit have been proposed as an explanation for the "red supergiant problem," and these effects should also strongly impact the rate, frequency, and progenitor detections of different supernova subtypes and progenitor detections.

While this is a clear predicted consequence of stellar evolution, observational evidence for these more massive stars dying as yellow supergiants or Wolf–Rayet stars is limited. These stars are still expected to have high luminosities, and should be readily detectable in pre-explosion imaging; however, only a handful of yellow supergiant progenitor candidates have been detected for II-L and IIb supernovae

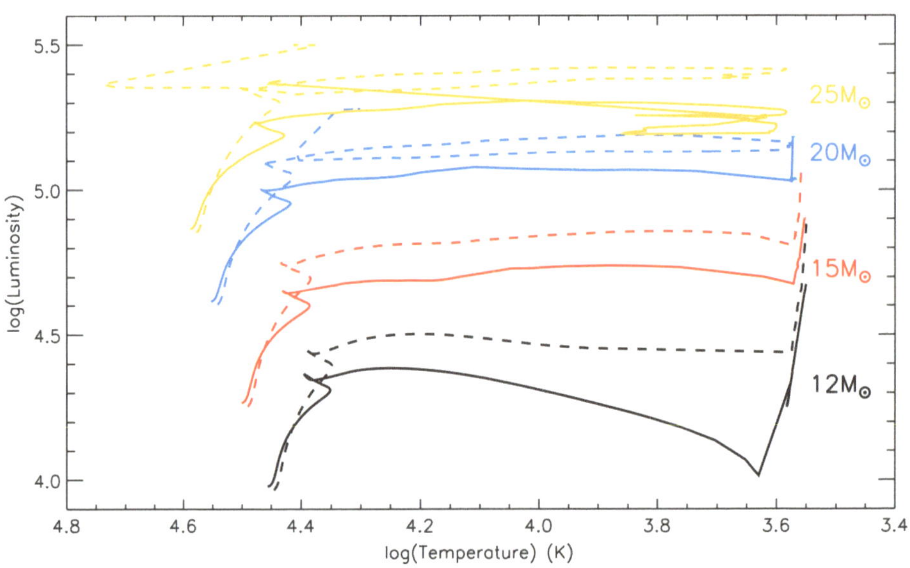

Figure 8.4. Comparison of the $12M_\odot$ (black), $15M_\odot$ (red), $20M_\odot$ (blue), and $25M_\odot$ (orange) Geneva evolutionary tracks at solar metallicity for non-rotating (solid lines) and rotating (dashed lines) stars, from Ekström et al. (2012). Note that for non-rotating stars the minimum terminal mass of a RSG is $\sim 20M_\odot$, while for rotating stars it drops to $15M_\odot$.

(see Smartt 2015). The most robust of these is the progenitor of SN 2011dh (Van Dyk et al. 2011; 2013; Maund et al. 2011). Its multi-wavelength detection was well-fit with the spectral energy distribution of a single ~6000 K yellow supergiant progenitor and associated with a Type IIb supernova—that is, a supernova that begins with the spectral signature of a typical Type II supernova with clear Balmer lines and later evolves to resemble the low-hydrogen or hydrogen-free Type Ib supernovae (see also Section 5.3.2). The only progenitor detection for a Type Ib supernova—which, along with the He-poor Type Ic subclass is presumed to come from Wolf–Rayet stars—is the blue progenitor of iPTF13bvn, which has been interpreted both as a potential low-mass Wolf–Rayet star (e.g., Cao et al. 2013; Groh et al. 2013) and as a lower-mass helium star (Bersten et al. 2014; Folatelli et al. 2016; Eldridge & Maund 2016).

Finally, SN 1987A and SN 1993J clearly illustrate the havoc that binary interactions can introduce to the predicted progenitor properties of a supernova. As discussed in Chapter 5 (and as illustrated in the binary evolutionary tracks included in Figure 8.3), mass loss from binary interactions can strongly impact the pre-explosion properties of an supernova progenitor. In the case of SN 1987A it is believed that a binary merger ejected a large quantity of mass, producing a blue supergiant progenitor in the years prior to core collapse (see Hillebrandt & Meyer 1989; Podsiadlowski et al. 1992; and Section 5.3.2). Similarly, pre-explosion imaging of the SN 1993J progenitor revealed a complex spectral energy distribution that could not be satisfactorily fit by a single star; later post-explosion imaging revealed a surviving blue B-type star and suggested that binary-induced mass loss prior to core collapse could have helped contribute to the mass stripping of an RSG and led to the supernova spectrum that classified SN 1993J as a Type IIb (see Fox et al. 2014; Section 5.3.2).

While these supernovae do not directly follow an RSG stage, the prior behavior of their progenitors during the RSG phase plays a key role. Phenomena such as mass-loss processes and dust production during the RSG phase (Chapter 4); the frequency of case B mass-transfer, mergers, and other binary interactions in RSGs (Chapter 5); and the relationship between RSG properties and the metallicity of their natal environments (Chapter 6) all impact their terminal evolution. Understanding RSGs is therefore critical not just for Type II-P supernovae, but for all core-collapse supernovae whose progenitors may have spent time as RSGs.

References

Abbott, B. P., Abbott, R., Abbott, T. D., et al. 2017, PhRvL, **118, 1101**

Abbott, B. P., Abbott, R., Abbott, T. D., et al. 2016a, PhRvL, **116, 1102**

Abbott, B. P., Abbott, R., Abbott, T. D., et al. 2016b, PhRvL, **116, 1103**

Adams, S. M., Abbott, R., Abbott, T. D., et al. 2017, MNRAS, **468, 4968**

Beasor, E. R. & Davies, B. 2016, MNRAS, **462, 1269**

Belczynski, K., Holz, D. E., Bulik, T., & O'Shaughnessy, R. 2016, Natur, **534, 512**

Bersten, M. C., Benvenuto, O. G., Folatelli, G., et al. 2014, AJ, **148, 68**

Cao, Y., Kasliwal, M. M., Arcavi, I., et al. 2013, ApJL, **775, L7**

Dessart, L., Livne, E., & Waldman, R. 2010, MNRAS, **408, 827**

Dessart, L. & Hillier, D. J. 2011, MNRAS, **410, 1739**
Ekström S., Georgy, C., Eggenberger, P., et al. 2012, A&A, **537, 146**
Eldridge, J. J., & Maund, J. R. 2016, MNRAS, **461, L117**
Eldridge, J. J., & Stanway, E. R. 2016, MNRAS, **462, 3302**
Filippenko, A. V. 1997, ARA&A, **35, 309**
Folatelli, G., Van Dyk, S. D., Kuncarayakti, H., et al. 2016, ApJL, **825, L22**
Fox, O. D., Azalee Bostroem, K., Van Dyk, S. D., et al. 2014, ApJ, **790, 17**
Gerke, J. R., Kochanek, C. S., & Stanek, K. Z. 2015, MNRAS, **450, 3289**
Groh, J. Georgy, C., & Ekström, S. 2013, A&A, **558, L1**
Hillebrandt, W., & Meyer, F. 1989, A&A, **219, L3**
Jennings, Z. G., Williams, B. F., Murphy, J. W., et al. 2014, ApJ, **795, 170**
Jerkstrand, A., Fransson, C., Maguire, K., et al. 2012, A&A, **546, 28**
Kochanek, C. S., Khan, R., & Dai, X. 2012, ApJ, **759, 20**
Levesque, E. M., Leitherer, C., Ekstrom, S., Meynet, G., & Schaerer, D. 2012, ApJ, **751, 67**
Levesque, E. M., Stringfellow, G. S., Ginsburg, A. G., Bally, J., & Keeney, B. A. 2014, AJ, **147, 23**
Lovegrove, E., & Woosley, S. E. 2013 ApJ, **769, 109**
Maguire, K., Di Carlo, E., & Smartt, S. J. 2010, MNRAS, **404, 981**
Mattila, S., Smartt, S., Maund, J., Benetti, S., & Ergon, M. 2010, e-print (arXiv:1011.5494)
Maund, J. R., Fraser, M., Ergon, M., et al. 2011, ApJ, **739, 37**
Maund, J. R., Reilly, E., & Mattila, S. 2014, MNRAS, **438, 938**
Maund, J. R. 2017, MNRAS, **469, 2202**
Ofek, E. O., Sullivan, M., Shaviv, N. J., et al. 2014, ApJ, **789, 104**
Pejcha, O., & Prieto, J. L. 2015, ApJ, **806, 225**
Pejcha, O., & Thompson, T. A. 2015, ApJ, **801, 90**
Podsiadlowski, P., Joss, P. C., & Hsu, J. J. L. 1992, ApJ, **391, 246**
Rabinak, I., & Waxman, E. 2011, ApJ, **728, 63**
Shiode, J. H., & Quataert, E. 2014, ApJ, **780, 96**
Smartt, S. J., Eldridge, J. J., Crockett, R. M., & Maund, J. R. 2009, MNRAS, **395, 1409**
Smartt, S. 2015, PASA, **32, 16**
Smith, N., Mauerhan, J. C., Cenko, S. B., et al. 2015, MNRAS, **449, 1876**
Sukhbold, T., & Woosley, S. E. 2014, ApJ, **783, 10**
Sukhbold, T., Ertl, T., Woosley, S. E., Brown, J. M., & Janka, H.-T. 2016, ApJ, **821, 38**
Ugliano, M., Janka, H.-T., Marek, A., & Arcones, A. 2012, ApJ, **757, 69**
Van Dyk S. D., Li, W., Cenko, S. B., et al. 2011, ApJ, **741, 28**
Van Dyk S. D., Davidge, T. J., Elias-Rosa, N., et al. 2012a, ApJ, **143, 19**
Van Dyk S. D., Cenko, S. B., Poznanski, D., et al. 2012b, ApJ, **756, 131**
Van Dyk S. D., Zheng, W., Clubb, K. I., et al. 2013, ApJ, **772, 32**
Walmswell, J. J. & Eldridge, J. J. 2012, MNRAS, **419, 2054**
Williams, B. F., Peterson, S., Murphy, J., et al. 2014, ApJ, **791, 105**
Woosley, S. E. 2016, ApJ, **824, 10**
Yaron, O., Perley, D. A., Gal-Yam, A., et al. 2017, NatPh, **13, 510**
Yoon, S.-C. & Cantiello, M. 2010, ApJ, **717, 62**

Chapter 9

The Future of Red Supergiant Research

Why write a book on RSGs now? As noted in the opening chapter, RSGs are crucial objects if we wish to understand stellar evolution, massive stellar populations, dust production and enrichment, and supernovae. Many of the fields connected with RSGs—variables and transients, extragalactic massive star populations, supernovae, compact object formation, and IR astronomy—are poised to experience a renaissance in the coming years. A host of groundbreaking new observational facilities will be coming online in the late 2010's and through the 2020's and will be met with an unprecented degree of computing power optimized for doing astronomy in the era of big data. Two substantial upcoming areas of growth in astronomy—and two areas where RSGs are vitally important—are time domain astronomy and the push of stellar observations out to impressive new distances.

9.1 Time Domain Astronomy

In just a few short years, the Large Synoptic Survey Telescope (LSST) will expand time domain astronomy to an unpredecented scale and usher in a new era of transient astronomy. Scheduled for science first light in 2021, this facility will image a tremendous swath (18,000 deg^2) of the southern sky in its deep-wide-fast survey. In a single year of operation it will yield a jaw-dropping 5.5 petabytes of raw data and lightcurves for ~300,000 new transients. The application to core-collapse supernovae is clear—LSST will provide an orders-of-magnitude increase over the number of core-collapse events currently being detected—but these observations will also dramatically increase the detection rate and volume of data for variable stars, binaries, and non-terminal transient events.

RSGs are extremely compelling objects for time domain astronomy. Their most obvious direct application is in their role as supernova progenitors—the detection rate of Type II-P supernovae will reach a fever pitch with LSST, and combined with massive new imaging campaigns (see below), the incidence of matching supernovae

doi:10.1088/978-0-7503-1329-2ch9

with detections in pre-explosion imaging should increase rapidly (see Section 8.1). Ideally this will allow us to develop a statistically robust sample of pre-explosion RSG detections, and to carry out larger-scale comparisons between progenitor and explosion properties in the Type II-P subset. LSST detections also make it possible to study the behavior of these supernovae at early times, to rapidly identify targets that should be followed up spectroscopically, and to monitor the evolution of the supernova lightcurve at a high (and, depending on the final survey design, multi-color) cadence. This will improve our ability to use the supernovae themselves to infer progenitor properties such as, for example, the interaction between the supernova shock and circumstellar material. LSST will also drastically increase our sample of other core-collapse supernovae, offering new information on the evolution of their progenitors and how the RSG phase might impact these stars' late-time evolution (Section 8.2).

Variable and non-terminal transient phenomena in Milky Way and even Local Group RSGs will also be observable with LSST. This includes monitoring photometric variability in an enormous sample of RSGs that extends down to the lower mass limit of the population, and evaluating spectroscopic variability based on apparent color changes in the RSG sample with time (combined with identifying ideal spectroscopic follow-up candidates). As discussed in Chapter 7, the size, depth, and cadence of this sample will be extremely valuable for improving our understanding of the underlying physical phenomena that drive RSG variability and the role that they play in the stars' evolution. Southern RSGs that are extremely photometrically variable, such as the OH/IR stars (see Section 4.3) or the TŻO candidate HV 2112 (Section 5.3.3), will be excellent targets for long-term monitoring. LSST will also offer a new avenue for studying sporadic mass-loss behavior in RSGs. Adams et al. (2017) identified an optical transient in the lightcurve of N6946-BH1 that preceded its presumed direct collapse to a black hole; however, there has never been a dedicated study of non-terminal transient behavior in RSGs, despite strong evidence that these stars do experience episodic or even eruptive mass loss. LSST lightcurves of the nearby RSG population will make this possible. Finally, the field of view and cadence of LSST will make it a powerful tool for detecting additional "vanishing" RSGs and other transient phenomena in these stars.

Moving beyond LSST, the role of the mass-losing RSG phase in the evolution and core-collapse death of massive stars has recently taken on a new importance. The detection of gravitational waves from a coalescing black hole binary by the Advanced Laser Interformeteter Gravitational-Wave Observatory (AdvLIGO) was announced in February of 2016 (Abbott et al. 2016a, 2016b), immediately establishing a entirely new avenue for studying the end products of massive star evolution. The masses of these black holes immediately raised new questions—the detections suggest a surprisingly high frequency of relatively massive stellar black holes ($\sim$10–40$M_\odot$; see also Abbott et al. 2017a). More recently, the detection in August of 2017 of gravitational waves and a short-duration gamma-ray burst from a neutron star merger has now moved gravitational wave studies of compact stellar remnants into the multi-messenger era (e.g., Abbott et al. 2017b, 2017c). With plans for future gravitational wave observatories it is expected that this new frontier will extend to the detection of other predicted gravitational wave sources, including core-collapse supernovae (Kotake et al. 2006, Abadie et al. 2012) and even weaker persistent (rather than transient) gravitational wave sources such as axially asymmetric rotating neutron stars (Abadie et al. 2010; Aasi et al. 2014, 2015; see Figure 9.1). Similar surprises arising from the eventual detection of these events can only be guessed at. Regardless, the rates of interacting and merging binary systems, the mass loss of compact object progenitors during their lifetimes, and the physics of core collapse are all key ingredients in the production of neutron stars and stellar mass black holes, and in each case the RSG phase plays a central role.

9.2 Beyond the Local Group

RSGs remain some of our most powerful "magnifying glasses" for studying extragalactic populations of massive stars and testing the predictions of theoretical evolutionary models. As noted in Chapter 6, RSGs in other galaxies give us the opportunity to observationally probe a variety of different host environments, testing how parameters such as metallicity impact the physical properties, ages, and relative populations of the massive stars.

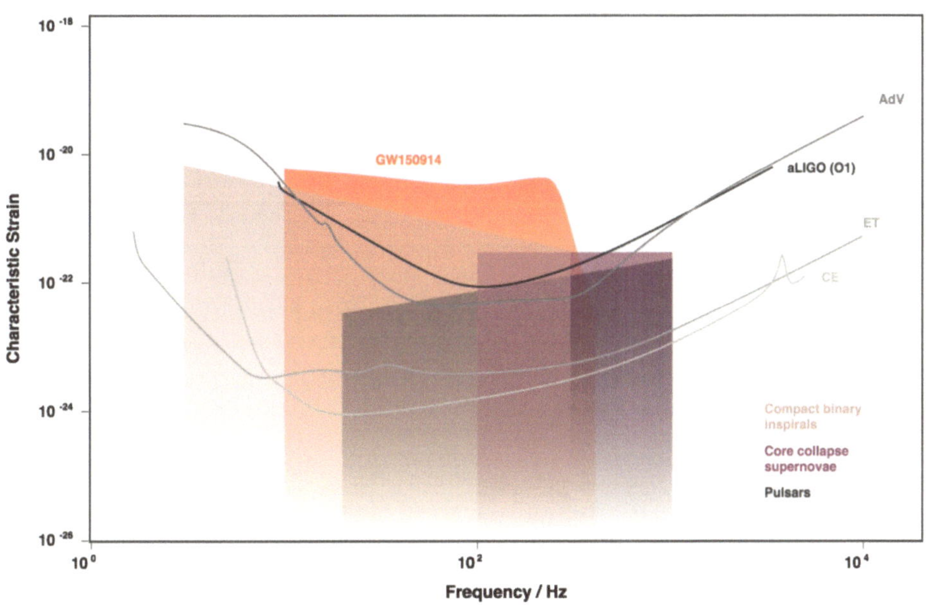

Figure 9.1. Generated using Gravitational Wave Sensitivity Curve Plotter by Christopher Moore, Robert Cole, and Christopher Berry, which is licensed under a Creative Commons Attribution-NonCommercial-ShareAlike 3.0 Unported License (http://rhcole.com/apps/GWplotter/). Adapted from Moore et al. (2015); frequency vs. characteristic strain sensitivity for a variety of RSG-related phenomena expected to produce gravitational waves, as compared to the measured sensitivity of Advanced LIGO's first observing run (aLIGO O1, black) and the planned sensitivity of several second- and third-generation gravitational wave observatories, including: Advanced VIRGO (AdV, dark gray), the Einstein Telescope (ET, gray), and the Cosmic Explorer (CE, light gray).

Figure 9.2. Artists' renditions of the Giant Magellan Telescope (left; courtesy of Giant Magellan Telescope-GMTO Corporation), Thirty Meter Telescope (center; courtesy of TMT International Observatory / T. Pyle (Caltech / IPAC)), and European Extremely Large Telescope (right; courtesy of ESO under a Creative Commons Attribution 4.0 International License), all $\gtrsim$24 m class telescopes scheduled for first light in the coming decade. Images are from http://www.gmto.org/wp-content/uploads/GMT-2015-sideview1%20large.jpg (left), http://www.tmt.org/sites/default/files/imagecache/Large/images/gallery/top%20view%20of%20tmt%20complex.jpg (center), and http://www.eelt.org.uk/img/elt-2011-02-cc.jpg (right).

RSGs can also potentially serve as diagnostics of these galaxies in their own right. Recent work has highlighted the potential utility of near-IR RSG spectra as metallicity diagnostics (e.g., Davies et al. 2010, 2015; Gazak et al. 2015, Patrick et al. 2015, 2016, 2017). Abundances determined from stellar spectra offer a compelling and robust alternative to those determined from other methods, such as strong emission line ratio diagnostics that measure the gas phase metallicity (which often show disagreements between different methods; see Kewley & Ellison 2008). Metallicities from blue supergiant spectra have previously been proven to be successful, but the extension to RSGs and analyses in the near-IR is particularly valuable for future observations.

The Extremely Large Telescopes (Figure 8.2) are all scheduled for first light in the 2020's, and all have multi-object spectrographs, IR spectrographs, and multi-conjugate or laser tomography adaptive optics systems planned as first light instruments. This combination of unprecedented light-gathering capability and adaptive optics is ideal for studying extragalactic RSGs in distant galaxies. Evans et al. (2011) have demonstrated that spectra of $J \sim 24$ mag RSGs will be possible out to ~50 Mpc, with the limiting factor being crowding rather than depth and the capabilities of the telescope itself (see Gazak et al. 2015). This opens up an astonishing new volume of observable spectra for individual massive stars. In addition to eventually yielding a rich suite of stellar metallicities for star-forming galaxies, we will be able to determine detailed physical properties for RSGs across an even broader range of different environments and compare these to predictions of stellar evolutionary theory. This is particularly compelling when considering the possibility of studying RSGs in extremely metal-poor environments, which can serve as local analogs of the environments in the early universe and test models of how the first stars evolved and died.

In addition, the James Webb Space Telescope (JWST) and Wide Field Infrared Space Telescope (WFIRST) are both expected to launch in the next decade. At the time of publication for this book, JWST is expected to launch in the spring of 2019 and will serve as the powerhouse space telescope for the next generation of astronomers. WFIRST is expected to launch in the mid-2020's; its wide-field imager will possess a tremendous 0.28 deg^2 field of view, complementing the incredible depth and resolving power of JWST. Both of these facilities will be excellent for observations of RSGs. Between the depth of JWST and the wide field of WFIRST, near-IR imaging of stellar populations in nearby galaxies will be ubiquitous, deep, and ideal for identifying individual RSGs. This is particularly valuable when combined with LSST and the potential for pre-explosion imaging of SN progenitors. The near-IR focus of these telescopes is also ideal for further exploring RSGs (and other massive stars) at low Galactic latitudes (see Section 6.2).

9.3 NGC 6946: An Example of RSG Astronomy in the 2020s

As an example of what will be possible with our observational capabilities in the coming decade, we can consider the example of RSGs in the nearby galaxy NGC 6946. NGC 6946 is ~5.5 Mpc away and is a spectacular face-on spiral galaxy with a substantial young massive star population. It is sometimes nicknamed the "Fireworks Galaxy" due to its incredible history as a core-collapse supernova host. NGC 6946 hosted 10 core-collapse supernovae in the 100-year period between 1917 and 2017—all have been designated as Type II-P supernovae except for the earliest events, which occurred when detailed lightcurve monitoring for this classification was not yet available. It is also the host of the possible disappearing RSG N6946-BH1 (Adams et al. 2017). This suggests that NGC 6946 has an immense population of RSGs nearing the ends of their lives (see Figure 9.3).

With a high northern declination, NGC 6946 will unfortunately *not* be observed with LSST. However, many similar transient surveys are currently under development, and it is easy to imagine what deep and high-cadence monitoring of NGC 6946 will produce: extremely rapid detection and spectroscopic follow-up of new core-collapse supernovae will be possible, as well as potential detection of pre-supernova outbursts and variability from RSGs. JWST will be able to obtain deep IR imaging of star-forming regions in NGC 6946, making it possible to study the RSG population (and surrounding massive star population) in detail and identify individual RSG members directly through optical+IR selection criteria; mid-IR data will also be particularly useful for identifying dust-enshrouded RSGs. With its wide field of view, WFIRST will be able to image the entire galaxy in only a handful of pointings, making it an invaluable resource for pre-explosion imaging of this galaxy if its current supernova rate holds. Finally, with RSG members of NGC 6946 identified, the Thirty Meter Telescope will be capable of obtaining spectroscopy of individual RSGs in this galaxy. With these spectra we will be able to determine precise local metallicities within the host, measure the stars' physical properties, and eventually assemble a large sample of well-understood extragalactic RSGs that can be considered future supernova progenitor candidates. NGC 6946 will unarguably be a hotbed of new RSG-related results in the near future.

Perhaps most exciting is the question of what we do not yet expect to discover with this suite of new facilities. While expanding our observational capabilities never ceases to surprise us with the new and unexpected, it seems reasonable to predict that some of the most exciting discoveries in the coming decades will be closely tied to

Figure 9.3. Courtesy of NASA. Color-composite image of NGC 6946, the "Fireworks Galaxy" NGC 6946. The locations of ten core-collapse Type II/II-P supernovae (yellow) and the candidate disappearing RSG N6946-BH1 (red) are indicated.

massive stars, stellar evolution, and supernovae, all areas where RSGs play a central role. Compiling our current body of knowledge of RSGs, and prioritizing continued efforts to study these stars in the coming years, will equip the astronomical community with the tools we need to take full advantage of the frontiers opened up by the next generation of observational facilities, allowing us to immediately begin probing the new corners of the universe that they reveal.

References

Abbott B. P., Abbott, R., Abbott, T. D., et al. 2017a, PhRvL, **118, 221101**
Abbott, B. P., Abbott, R., Abbott, T. D., et al. 2017b, ApJL, **848, 12**
Abbott, B. P., Abbott, R., Abbott, T. D., et al. 2017c, ApJL, **848, 13**
Abbott B. P., Abbott, R., Abbott, T. D., et al. 2016a, PhRvL, **116, 1102**
Abbott B. P., Abbott, R., Abbott, T. D., et al. 2016b, PhRvL, **116, 1103**
Adams, S. M., Kochanek, C. S., Gerke, J. R., Stanek, K. Z., & Dai, X. 2017, MNRAS, **468, 4968**
Aasi, J., et al. 2014, RhRvD, **90, 062010**
Aasi, J., Abbott B. P., Abbott, R., et al. 2015, ApJ, **813, 39**
Abadie, J., Abbott B. P., Abbott, R., et al. 2010, ApJ, **722, 1504**
Abadie, J., Abbott, R., Abbott, T. D., et al. 2012, PhRvD D, **85, 122007**

Davies, B., Kudritzki, R.-P., & Figer, D. F. 2010, MNRAS, **07, 1203**
Davies, B., Kudritzki, R.-P., Gazak, Z., et al. 2015, ApJ, **806, 21**
Dong, S. & Stanek, K. 2017, ATel, **10372, 1**
Evans, C. J., Davies, B., Kudritzki, R.-P., et al. 2011, A&A, **527, 50**
Kewley, L. J., & Ellison, S. L. 2008, ApJ, **681, 1183**
Gazak, J. Z., Kudritzki, R., Evans, C., et al. 2015, ApJ, **805, 182**
Kotake, K., Sato, K., & Takahashi, K. 2006, RPPh, **69, 971**
Moore, C. J., Cole, R. H., & Berry, C. P. L. 2015, CQG, **32, 015014**
Patrick, L. R., Evans, C. J., Davies, B., et al. 2015, ApJ, **803, 14**
Patrick, L. R., Evans, C. J., Davies, B., et al. 2016, MNRAS, **458, 3968**
Patrick, L. R., Evans, C. J., Davies, B., et al. 2017, MNRAS, **468, 492**

www.ingramcontent.com/pod-product-compliance
Ingram Content Group UK Ltd.
Pitfield, Milton Keynes, MK11 3LW, UK
UKHW061829100926
8556IPUK00011B/932
* 9 7 8 0 7 5 0 3 1 3 3 0 8 *